EVOLUTION

GENETICS & EVOLUTION

EVOLUTION

The History of Life on Earth

RUSS HODGE

FOREWORD BY NADIA ROSENTHAL, PH.D.

Facts On File
An imprint of Infobase Publishing

This book is dedicated to the memory of my grandparents, E. J. and Mabel Evens and Irene Hodge, to my parents, Ed and Jo Hodge, and especially to my wife, Gabi, and my children—Jesper, Sharon, and Lisa—with love.

❧

EVOLUTION: The History of Life on Earth

Copyright © 2009 by Russ Hodge

Facts On File, Inc.
An imprint of Infobase Publishing
132 West 31st Street
New York NY 10001

Library of Congress Cataloging-in-Publication Data

Hodge, Russ, 1961–
    Evolution: the history of life on earth / Russ Hodge; foreword by Nadia Rosenthal.
        p. cm.
    Includes bibliographical references and index.
    ISBN-13: 978-0-8160-6679-7
    ISBN-10: 0-8160-6679-5
    1. Evolution (Biology)—History. I. Title.
    QH361.H635 2009
    576.8—dc22   2008029741

Facts On File books are available at special discounts when purchased in bulk quantities for businesses, associations, institutions, or sales promotions. Please call our Special Sales Department in New York at (212) 967-8800 or (800) 322-8755.

You can find Facts On File on the World Wide Web at http://www.factsonfile.com

Text design by Kerry Casey
Illustrations by Lucidity Information Design
Photo research by Elizabeth H. Oakes

Printed in the United States of America

Bang FOF 10 9 8 7 6 5 4 3 2

This book is printed on acid-free paper.

"I say that it touches a man that his blood is sea water and his tears are salt, that the seed of his loins is scarcely different from the same cells in a seaweed, and that of stuff like his bones coral is made. I say that the physical and biologic law lies down with him, and wakes when a child stirs in the womb, and that the sap in a tree, uprushing in the spring, and the smell of the loam, where the bacteria bestir themselves in darkness, and the path of the sun in the heaven, these are facts of first importance to his mental conclusions, and that a man who goes in no consciousness of them is a drifter and a dreamer, without a home or any contact with reality."

—from *An Almanac for Moderns:*
A Daybook of Nature
by Donald Culross Peattie
copyright © 1935 (renewed 1963)
by Donald Culross Peattie

Contents

Foreword

It hits everyone differently, but the feeling of infectious curiosity is unmistakable. My own obsession with biology sprung, unexpectedly, out of an early passion for art. My epiphany at fifteen was sparked as much by the recurring themes in nature that I had been trying to capture in paint as by the biological information I was gobbling up at school, fed to me by an inspired teacher. I soon became convinced that studying the living world would satisfy my curiosity more than painting ever could.

Reading Russ Hodge's book reminds me of the excitement of those early days in my own discovery of evolution. It is a challenging, fascinating account: engaging narratives of the men and women with open eyes who struggled to reconcile their religious faith in a story that did not fit the facts they were uncovering in the fossils dug out from shales and cliffs of foreign shores, or even in their own backyards. Against a backdrop of our own relatively brief human history, the immense scale of geological time embedded in those rocks was a real conceptual impediment to these thinkers. To grasp the gradual emergence of life-forms, they had to wrestle with the presumptions of the day, that organisms spontaneously could appear fully formed, rather than evolve from a previous version. These people were courageous, real explorers of the unknown and reading their stories reminds me of the reasons I became obsessed with biology in the first place. Reading about them can help today's students experience for themselves the excitement of scientific exploration, and to appreciate how profoundly the work of yesterday's scientists affects the quality of our lives today. In trying to predict what new insights we will gain through biology in the course our own lifetimes, it's helpful to put oneself in the shoes of someone who did not yet know what DNA was, yet somehow imagined the ways genes might work from pure observation of his natural surroundings. I still thrill when

I read of Wallace's solitary epiphany of evolutionary process in the Pacific islands, and still wince at the principles of evolution wrongly applied to human racial segregation. These are the tales this book tells.

The story of evolutionary discovery shows us as members of modern society how important it is to gain at least a basic grasp of scientific knowledge. The original concept of natural selection, so controversial that both of its founders hesitated for years to publish it, could have been readily tested with access to today's information and technology we now take for granted. What Darwin wouldn't have given for one of our state-of-the-art gene sequencing machines that provides us with an entire human DNA sequence in a day! The scientific opportunities in front of us have never been more exciting, but also raise controversial questions. Our own personal genetic data will soon be available, giving us a chance to chart our private evolution in a way never before possible. But more information can always be misused, and as a society it's critical that we engage with current scientific issues that will shape the future. To deal with the challenges of today's science, we must be armed with at least a general understanding of the concepts underlying the advances that brought us to this point in history.

These are the sensible reasons why studying the story of evolution matters. But the real reason is less practical. As the physicist Brian Greene describes it, science gives us a new perspective, and it allows us to travel from confusion to understanding in a manner that's precise, predictive, and reliable. To him, doing physics is a transformation that actually feels empowering and emotional. As a student I was lucky to have a teacher who made me feel that way about biology before I knew very much about it. To be able to think through and grasp the explanations that reveal how life formed on earth is a precious human experience that you do not have to be a scientist to enjoy.

Evolution is often viewed as an isolated topic that sometimes shows up in the news as a controversial challenge to biblical accounts of creation. But the study of evolution actually provides us with a new dimension of experience, giving us a language to connect our own history as a species to our present lives, and enriching our encounters with the natural world around us. *Evolution: The History of Life on Earth* will open those new dimensions and teach you the basics of the language while taking you on a great human adventure. It captures the drama of the struggle with religious belief, and that epiphany of insight that is at the core of scientific discovery. It will convince you that evolution matters.

—Nadia Rosenthal, Ph.D.
Head of Outstation,
European Molecular Biology Laboratory
Rome, Italy

Preface

In laboratories, clinics, and companies around the world, an amazing revolution is taking place in our understanding of life. It will dramatically change the way medicine is practiced and have other effects on nearly everyone alive today. This revolution makes the news nearly every day, but the headlines often seem mysterious and scary. Discoveries are being made at such a dizzying pace that even scientists, let alone the public, can barely keep up.

The six-volume Genetics and Evolution set aims to explain what is happening in biological research and put things into perspective for high-school students and the general public. The themes are the main fields of current research devoted to four volumes: *Evolution, The Molecules of Life, Genetic Engineering,* and *Developmental Biology.* A fifth volume is devoted to *Human Genetics,* and the sixth, *The Future of Genetics,* takes a look at how these sciences are likely to shape science and society in the future. The books aim to fill an important need by connecting the history of scientific ideas and methods to their impact on today's research. *Evolution,* for example, begins by explaining why a new theory of life was necessary in the 19th century. It goes on to show how the theory is helping create new animal models of human diseases and is shedding light on the genomes of humans, other animals, and plants.

Most of what is happening in the life sciences today can be traced back to a series of discoveries made in the mid-19th century. Evolution, cell biology, heredity, chemistry, embryology, and modern medicine were born during that era. At first these fields approached life from different points of view, using different methods. But they have steadily grown closer, and today they are all coming together in a view of life that stretches from single molecules to whole organisms, complex interactions between species, and the environment.

The meeting point of these traditions is the cell. Over the last 50 years biochemists have learned how DNA, RNA, and proteins carry out a complex dialogue with the environment to manage the cell's daily business and to build complex organisms. Medicine is also focusing on cells: Bacteria and viruses cause damage by invading cells and disrupting what is going on inside. Other diseases—such as cancer or Alzheimer's disease—arise from inherent defects in cells that we may soon learn to repair.

This is a change in orientation. Modern medicine arose when scientists learned to fight some of the worst infectious diseases with vaccines and drugs. This strategy has not worked with AIDS, malaria, and a range of other diseases because of their complexity and the way they infiltrate processes in cells. Curing such infectious diseases, cancer, and the health problems that arise from defective genes will require a new type of medicine based on a thorough understanding of how cells work and the development of new methods to manipulate what happens inside them.

Today's research is painting a picture of life that is much richer and more complex than anyone imagined just a few decades ago. Modern science has given us new insights into human nature that bring along a great many questions and many new responsibilities. Discoveries are being made at an amazing pace, but they usually concern tiny details of biochemistry or the functions of networks of molecules within cells that are hard to explain in headlines or short newspaper articles. So the communication gap between the worlds of research, schools, and the public is widening at the worst possible time. In the near future young people will be called on to make decisions—large political ones and very personal ones—about how science is practiced and how its findings are applied. Should there be limits on research into stem cells or other types of human cells? What kinds of diagnostic tests should be performed on embryos or children? How should information about a person's genes be used? How can privacy be protected in an age when everyone carries a readout of his or her personal genome on a memory card? These questions will be difficult to answer, and decisions should not be made without a good understanding of the issues.

I was largely unaware of this amazing scientific revolution until 12 years ago, when I was hired to create a public information office at one of the world's most renowned research laboratories. Since that time I have had the great privilege of working alongside some of today's greatest researchers, talking to them daily, writing about their work, and picking their brains about the world that today's science is creating. These books aim to share those experiences with the young people who will shape tomorrow's science and live in the world that it makes possible.

Acknowledgments

This book would not have been possible without the help of many people. First I want to thank the dozens of scientists with whom I have worked over the past 12 years, who have spent a great amount of time introducing me to the world of molecular biology. In particular I thank Volker Wiersdorff, Patricia Kahn, Eric Karsenti, Thomas Graf, Nadia Rosenthal, and Walter Birchmeier. My agent, Jodie Rhodes, was instrumental in planning and launching the project. Frank Darmstadt, executive editor at Facts On File, kept things on track and made great contributions to the quality of the text. Sincere thanks go as well to the Production and Art departments for their invaluable contributions. I am very grateful to Beth Oakes for locating the photographs for the entire set. Finally, I thank my family for all their support. That begins with my parents, Ed and Jo Hodge, who somehow figured out how to raise a young writer, and extends to my wife, Gabi, and children, Jesper, Sharon, and Lisa, who are still learning how to live with one.

Introduction

If each of us lived for tens of millions of years, the world would seem like a much different place. We would witness the birth and death of mountain ranges and the steady drift of continents; ice ages would pass like brief storms. New creatures would appear, undergo amazing transformations, and vanish again. Our short life spans offer only snapshots of these processes. But imagination and science give us the unique ability to take new perspectives. We can discover the height of a mountain without climbing it, using surveying instruments, a few basic facts, and the rules of geometry. We cannot go back in time, but experiments can be used to analyze the formation of geological layers and measure the age of the Earth. And we can take a similar approach to the study of life.

In the 19th century Charles Darwin, Alfred Russel Wallace, and their contemporaries were faced with a set of puzzles about life that had no scientific explanation. Why did the Earth hold so many species—thousands of types of beetles alone, many of which differed only in tiny details? Why did each continent have a unique set of animals, and what was the meaning of *fossils*? Why did the bones of a dolphin's fin look so similar to those in a human's hand? Why did different strata of rocks contain different fossils? Darwin and Wallace found answers by imagining what the Earth would look like over hundreds of millions of years.

Their solution, the theory of *evolution,* is based on three main principles—*heredity, variation,* and *natural selection*—that people have been aware of for a long time and can be observed anywhere, from exotic tropical rain forests to our own backyards. The huge breakthrough of Darwin and Wallace was to explain how these processes work together to produce new species. Evolution provided the first scientific system to investigate the origins and relationships of living creatures, and today it serves as a grand unifying theory that explains facts that cannot really be accounted for in any other way. The goal of this book is to demonstrate why the theory was necessary, to describe it as clearly as possible, to show how it has been received by society, and to explain the central role that it plays in today's science. Another theme of the book is the immense impact that evolution has had on society.

Most books on evolution focus on the *theory* itself, or the life and ideas of Charles Darwin, or the history of life on Earth. Few of them explain what evolutionary theory means in today's science, in fields such as modern medicine. Nearly all of the books that do are written for college students or specialists. There is a need for a book that shows how these aspects of the theory fit together, at a level that can be understood by high-school students and the general public. *Evolution* aims to fill that niche.

The first chapter sets the stage by describing how people understood the world in the early 19th century, showing some of the historical influences on the ideas of Darwin and Wallace, and explaining why a central theory was necessary to explain what scientists were discovering about life. Chapter 2 recounts how the personal experiences and observations of Darwin and Wallace led to their theories. The third chapter explains the theory itself and discusses its enormous impact on society. It was accepted quickly by most scientists, and it encouraged philosophers and industrialists to look at human culture in terms of evolution; some of them hoped to use it to change society. Most of these attempts failed because people misunderstood the theory, or used it as an excuse to promote political or racist agendas.

The birth of genetic science in the early 1900s and the discovery that genes were made of DNA in the 1950s opened a new era in evolutionary science. Chapter 4 describes how evolution and *genetics* began to flow together in the early 20th century, creating modern biology. Chapter 5 shows how the discovery of the double-helical structure of DNA instantly brought the science of evolution to the level of molecules and the chemistry of cells. The final chapter gives some examples of the crucial role that evolutionary theory has played in research in the age of *genomes.*

Very little of what biologists do today would make sense if life had not evolved in the way Darwin and Wallace proposed. Within science, the major aspects of the theory are no longer seriously debated. And whereas most people throughout the world, major religions, and even the Pope have embraced evolution, it is still a topic that divides society, in the United States more than in most other countries. It is possible to heal this rift, but the process has to start with a very clear understanding of what evolution means and what it does not. Evolution is easy to misunderstand, and then it can easily be misused to promote political or other agendas. History shows that using science in this way can lead down a very dangerous road. By showing where the theory came from, how it developed, and how it is driving modern science, this book aims to help clarify some of these issues.

1

A World without Evolution

In the middle of the 19th century Charles Darwin (1809–89) and Alfred Russel Wallace (1823–1913), living on opposite sides of the globe, simulta- neously proposed a com- pletely new theory of life on Earth that contradicted what nearly everyone had believed until then. The theory answered a number of questions that had puz- zled scientists for a long time. To understand evo- lution, it is helpful to un- derstand what people be- lieved before its discovery, and why previous views of life were neither adequate nor scientific.

Charles Darwin in 1869, 10 years after the publication of *On the Origin of Species* (*University of Brunel*)

For centuries the goal of most research in the West- ern world had been to sup- port the biblical account of the universe. The religious movement called *natural theology,* also known as *intelligent design,* studied the world and its *species* with the goal of proving the existence of God. Yet a completely new way of understanding the universe was developing, one based on careful observations, hypotheses, predictions, and experimental testing. The Bible did not offer a complete description of the world and there were intense debates about how it should be interpreted. Amateur scientists of the early 19th century believed that understanding the laws of nature might fill in some of the gaps. A theory of life was necessary for the same reasons that the Earth was no longer seen as the center of the universe and geologists had begun to suspect that the planet was very ancient. This chapter explains the origins of the idea of evolution and how it relates to the development of a scientific method of understanding the world.

CREATION, RELIGIONS, AND THE AGE OF THE EARTH

The world seems so mysterious and wonderful that people have always believed that gods created it. For the Australian aborigines, the world originated during an era called the "Dreamtime," when ancestral beings wandered the Earth and created humans out of bundled, shapeless bits of nature. A Scandinavian myth tells how animals and humans formed from the sweat of a sleeping god called Ymir.

Starting in about 700 B.C.E., the philosophers of ancient Greece began creating a new view of the world. They believed that laws that could be revealed through careful observation governed the universe. As well as proposing that the Earth was round and that substances were composed of tiny particles called atoms, they developed new ideas about the origins of life. After studying fossils, Anaximander (610–546 B.C.E.) claimed that all life arose from the sea, and that fishlike creatures moved onto land where they gave birth to humans and other animals. Greek science was still more descriptive and philosophical than experimental; it usually did not involve clear hypotheses that could be tested in a systematic way. But these early thinkers valued careful observation over dogmas and the theology of the past, and their goal was to reveal the underlying principles or laws that governed the natural world.

Judeo-Christian beliefs about the origins of life are recounted in Genesis, the first book of the Bible, which tradition attributes to the prophet Moses. This story dominated attitudes about nature and culture in the Western world and deeply influenced how society received the ideas of Darwin and Wallace.

The Bible states that God created the universe over a six-day period, in several steps. On the second day God created the "firmament" (the Earth), and on the third day dry land appeared, along with grasses and plants. On the fourth day God made the Sun, Moon, and stars, and on the fifth "every living creature that moves," with the exception of human beings (created on the sixth day). After each act of creation, in the words of the King James Version of the Bible, "God saw that it was good." This has been interpreted to mean that everything in creation was perfect, with no hint that natural forces might alter them. So any new theory that proposed changes in species contradicted a long religious tradition.

Evidence of the "goodness" of God's works could be found everywhere. Plants and animals depended on each other and seemed well-designed for their environments: Animals were fast enough to catch their prey; eagles' eyes were sharp enough to spot mice from high in the air. When it became legal and morally acceptable to study anatomy in the late Renaissance, physicians discovered a world within the human body far more intricate than the most complex man-made machine, one that surely must have been designed by an intelligent creator. This was stated eloquently by Colin Maclaurin (1698–1746), one of Sir Isaac Newton's students and a strong promoter of the scientific method at the University of Edinburgh, Scotland: "The plain argument for the existence of the Deity. . . . is from the evident contrivance and fitness of things for one another, which we meet with throughout all parts of the universe. . . . No person, for example, that knows the principles of optics and the structure of the eye, can believe that it was formed without skill in that science." The thinkers of the 18th century had no other explanation for how these things could have come about.

Passages in the Bible predicted that the world would last only 6,000 years, which was a powerful motivation for scholars to try to work out the exact date of creation—everyone wanted to be prepared for the end of the world. The early Christian scholars Julius Africanus (writing in 221 C.E.) and Bishop Eusebius of Caesaria (about 80 years later) based their calculations on biblical passages giving the ages of the first man and all of his descendents down to Moses. They arrived at a date of about 4,000 years before the birth of Christ. In 1650, the Calvinist Bishop James Ussher published a new set of calculations in a 2,000-page book called *Annals of the Old Testament, Deduced from the First Origins of the World.* He concluded that Creation began at nightfall before Sunday, October 23, 4004 B.C.E. Many other scholars followed his example and made chronologies based on the Bible. Even the great physicist Sir Isaac Newton (1643–1727), who had strong mystical beliefs in ancient prophecies, made a chronology that placed creation at about the same time.

Knowing the age of the planet was essential to the development of a science of life. A 6,000-year-old Earth gives very little time for change. When geologists such as James Hutton (1726–97) began to suspect that the Earth might be thousands or millions of times older than this, they opened the door to new ways of thinking about the origins of species. Like mountain ranges, they might have arisen through a gradual process, following laws that were still at work in the natural world.

ASTRONOMY AND PHYSICS POSE MAJOR CHALLENGES TO CHURCH DOCTRINES

The careful study of the sky began thousands of years ago as sailors learned to navigate using patterns of stars as reference points. Modern astronomy traces its roots back to about 1500, when the Polish scientist Nicolaus Copernicus (1473–1543) proposed a radical new theory that the Earth and other planets of our solar system orbited the Sun. In doing so, he moved the Earth out of the center of creation, just as evolution would later take away humans' central status in the living world. Martin Luther said of Copernicus that "the fool wishes to reverse the entire science of astronomy! But as sacred Scripture tells us, Joshua commanded the Sun to stand still, and not the Earth!"

In itself a revolving Earth might not have been a serious issue for the church even though it contradicted Scripture. Keen observers of nature had already discovered natural phenomena that seemed to contradict the Bible—for example, the book of Leviticus claims that locusts and grasshoppers have four legs, although they are insects with six. In the Middle Ages, scholars rediscovered the writings of the Greek philosopher Aristotle (384–322 B.C.E.) which contradicted many doctrines of the church. But Aristotle's belief in a single powerful god could be useful, so Thomas Aquinas (1225–74) and other theologians managed to import many of his ideas into Christianity. They also left things out. Aristotle did not, for example, believe in creation—he thought that the universe had existed forever. He also believed that existing species could become new ones over time.

Several of Aristotle's ideas about astronomy became recognized as official Church doctrines. He claimed that nothing in the heavens farther away than the Moon could ever change, and that the stars were embedded in a set of 22 crystalline spheres that rotated above the Earth in complex patterns. He concluded that the Earth had a spherical shape, because different stars could be seen if one traveled to the north or south. But he also believed that it occupied the center of the universe. He explains his reasoning in the book *On the Heavens,* written in 350 B.C.E.: "Any body endowed with weight, of whatever size, moves towards the center. Clearly it will not stop when its edge touches the center. The greater quantity must prevail until the body's center occupies the center. . . . Now it makes no difference whether we apply this to a clod or common fragment of earth or to the Earth as a whole. . . . Therefore Earth in motion, whether in a mass or in fragments, necessarily continues to

Portrait of Galileo (*Justus Sustermans, 1597–1681*)

move until it occupies the center equally every way, the less being forced to equalize itself by the greater owing to the forward drive of the impulse."

While the system of spheres seemed to account for the behavior of most of the night sky, it offered no explanation for comets or two supernovae (exploding stars) that suddenly appeared in the late 16th and early 17th centuries. Careful measurements showed that these strange objects lay far beyond the Moon. Another of Aristotle's doctrines stated that things could only orbit the Earth, not any other heavenly body—which would cause a problem when Galileo Galilei (1564–1642) plotted the orbits of Jupiter's moons.

As a professor at the university in Padua, Italy, Galileo taught that human observation and reason should be trusted over scriptures and ancient authorities when trying to answer certain questions about the world. The approach disturbed not only leaders of the church, but also Galileo's fellow professors. Galileo complained about his colleagues in a letter to fellow astronomer Johannes Kepler: "What do you think of the foremost philosophers of this University? In spite of my oft-repeated efforts and invitations, they have refused, with the obstinacy of a glutted adder, to look at the planets or Moon or my telescope." Scholars and clerics had no interest in the telescope because they assumed that nothing in human experience could possibly contradict the Scriptures; if this seemed to happen, it merely proved that human senses and understanding were flawed, or that Satan was using reason to lead people astray. The same arguments are used today to ignore discoveries in biology.

Galileo did not intend to call God's existence into question; he did not believe that science could contradict faith, but rather that studying nature would increase people's appreciation of God's work. He strongly felt, however, that the Church should update its views of nature. His findings and experiments revealed "natural laws" that could be described through mathematics. He might have received a fair hearing but for a terrible political blunder; in his book *A Dialogue on the Two World Systems,* he ridiculed the Church's point of view rather than approaching it with respect. Although the Pope had been a good friend of Galileo when the two men were younger, Urban VIII could not tolerate a direct insult, and the astronomer was sentenced to house arrest.

Newton followed up Galileo's work with mathematical descriptions of the laws of mechanics. His explanation of gravity has been called the discovery of the first "natural law." Newton's work established a method for the investigation of the world that forms the basis of all modern sciences. Observations should yield hypotheses; these can be used to make predictions that can be tested in experiments by anyone. If the results do not uphold a theory, it must be discarded. Other scientists would apply this method to geology and finally to the origins of life. Like Galileo, Newton saw no contradiction between science and religion as he laid the groundwork for modern science. He was a deeply religious man who strongly believed in

God and creation. In fact, most of his writings are devoted to alchemy, mysticism, and religion.

One of Newton's principles states that in a vacuum, an object set in motion continues to travel in a straight line forever, unless disturbed by another force. This is vital to understanding gravity's effect on the planets, and it influenced how people thought about God's relationship to the world. The creator did not have to constantly intervene to make sure that the universe functioned: Instead, he might have put into place the initial state of the world and laws to keep it running.

The Frenchman René Descartes (1596–1650) also made major contributions to the development of the scientific method. Descartes was willing to call everything into question, building a logical system of philosophy that supported the evidence of the senses over authority. He pondered what the world would be like if the smallest parts of our bodies (meaning something like atoms) behaved according to Newton's laws: If the positions and velocities of all the particles in the universe at any given point in time were known, as well as the forces acting on them, one might be able to predict everything in the future. Yet this did not mean that humans were simply machines; the soul and the body were separate things, he claimed, and an eternal soul could not arise from simpler components. It could only have been created by something more perfect than itself, and Descartes believed this proved that an eternal, all-powerful deity had to exist.

GEOLOGISTS POSTULATE AN ANCIENT EARTH

Mountains and landscapes seem permanent until a monstrous wave carries away a coastline, a volcano spouts fire, or cities are buried under thick layers of ash and stone. One of Plato's dialogues describes how the fabled city of Atlantis sank beneath the sea. Such terrible events imprint themselves into the memory of a culture, spawning superstitions and legends. In the 18th century, Newton's scientific method inspired a new generation of scientists to observe the world beneath their feet as closely as astronomers had scanned the skies.

It is probably no great surprise that the University of Edinburgh, in Scotland, became a center for geology. Edinburgh sits on the rim of an ancient volcano, so evidence of the powers that formed the Earth were as close as a glance out the classroom window. Colin Maclaurin, a young student of Newton's, obtained an influential teaching position there thanks to Newton's recommendation. In his classes he strongly advocated the scientific method of careful observations, theory building, and experimentation. Students were encouraged to try the method on everything: stars, chemistry, farming, and the Earth itself. Edinburgh became one of the most progressive universities of its day, with excellent faculties of medicine and science. Charles Darwin first attended the university there, and his older brother, his father, and his grandfather Erasmus, one of the most radical and brilliant thinkers of his day, all got their degrees there.

James Hutton enrolled at the university and was immediately initiated into the scientific method of Newton and Maclaurin. He wrote: "We must not allow ourselves ever to reason without proper data, or to fabricate a system of apparent wisdom in the folly of a hypothetical delusion." Hutton became a brilliant agriculturalist who developed methods that improved British farming while cultivating a passionate hobby, investigating the makeup of the Earth. His interest in chemistry and minerals were a great help as he studied the composition of rocks and thought about how the planet had formed.

The Earth's outer crust contains many thin layers of rock, or strata, stacked on top of each other. They can be seen along any stretch of highway where rock has been blasted away. Strata come in a particular order and tell a story, like the pages of a book lying facedown on a table: The deepest layers are the oldest, laid down the earliest, and higher layers represent successively more recent eras. This idea was first proposed by Niels Stensen (1638–86), a Danish physician and world traveler. Working in Florence for the Medici family, the patrons of Galileo a few decades earlier, Stensen (also called Nicolaus Steno) classified rocks into several types based on the processes that created them. His masterpiece, *De solido intra solidum naturalizer contento dissertationis prodromus,* or *Prodromus* for short, written in 1669, introduced new principles of geology, such as the idea that strata were originally lain down flat. The fact that some now aimed upward or appeared broken was due to later movements of the Earth. This had implications for time because it implied that different parts of a single stratum, even if spread over a large area, formed at the same time. Stensen concluded that strange triangular objects found in some of the strata were fossilized sharks' teeth. This led him to propose that strata contained a record of different eras of life.

A hundred years later, when James Hutton enrolled at the University of Edinburgh, Stensen's ideas were still not commonly accepted. If the world had sprung whole from the mind of a Creator in an instant, there was no pressing need to examine the Earth's crust for signs that it had been formed by a dynamic process. Except for a few notable events, such as volcanic eruptions, earthquakes, and the great flood of the Bible, scientists assumed that the world had always been close to the way they knew it now.

Hutton carefully studied the strata and the types of rocks he encountered while on walking tours through much of Scotland and England; the patterns he found made him begin to think seriously about time. Strata could be used as a great clock. Knowing how long it took each layer to form would provide a scientific means of dating the Earth.

Most rocks were sedimentary; erosion brought loose material into lakes or oceans, where it filtered to the bottom and hardened over time through heat or pressure. This usually happened to large, flat layers, which explained why Hutton found the same strata spread out over a large region, in the same

(opposite page) By the mid-19th century geologists had worked out the major geological periods in the history of the Earth and assembled diagrams like this one showing strata in their historical order. Each layer contained a unique set of fossils, which helped biologists understand how the types and forms of life had changed over immense stretches of geological time.

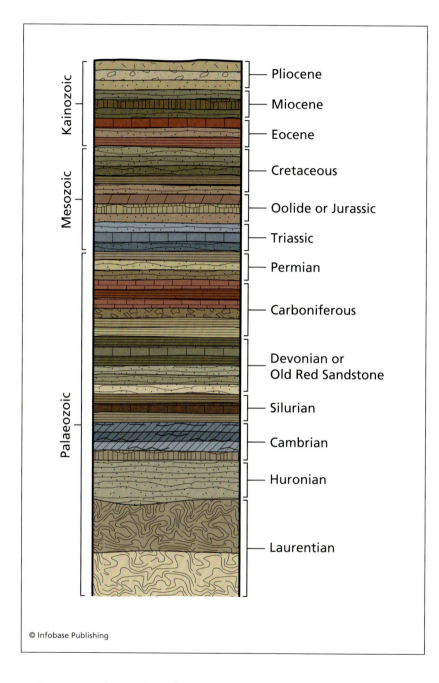

Pliocene

Miocene

Eocene

Cretaceous

Oolide or Jurassic

Triassic

Permian

Carboniferous

Devonian or
Old Red Sandstone

Silurian

Cambrian

Huronian

Laurentian

Kainozoic

Mesozoic

Palaeozoic

© Infobase Publishing

order. Yet a layer found on the surface in one place might be found far up on the side of a cliff in another, demonstrating that the Earth's crust had undergone numerous major upheavals. To capture the whole story, it would be necessary to assemble a complete and logical record of the order of strata in Britain.

This immense project was undertaken by the British surveyor William Smith (1769–1839) and a group of assistants. Smith made enormous, wall-sized maps of Great Britain that he colored in painstakingly by hand, giving each type of rock its own color. As he carried out this work, Smith came across the same phenomenon that Stensen had seen—each layer contained its own unique collection of fossils. This meant that fossils could be used to identify and compare strata where only a few layers were exposed, or in ambiguous cases. It also raised a huge question: Why did each layer contain different fossils? The answer lay in the connection between geological and biological time. Life on Earth had undergone major changes.

A great problem now facing geology was the question of how rocks lying at the bottom of the sea had made their way hundreds or even thousands of feet into the air. Intent on fitting the new findings to the Bible, German scientist Abraham Werner (1749–1817) proposed that strata and different types of rocks were created as the waters receded after the biblical flood. This was the accepted explanation in Edinburgh and elsewhere, but Hutton knew that erosion alone could not explain everything he was observing. Throughout England and Scotland he had seen evidence of incredible forces that seemed to have been generated from below, forcing strata upwards, bending and cracking them. If the level of the sea had simply dropped, he argued, all the strata seen in mountains and cliffs would still run parallel to the ground.

Granite was a key to resolving the debate. Veins of this stone could be seen running through sedimentary rocks. Werner claimed that granite veins were created by fluids seeping downward and then mineralizing. During field trips throughout Scotland, Hutton became convinced that granite started off as an intensely hot liquid that then burst upward in veinlike patterns.

Hutton now had a set of ideas that could be used to begin linking strata to time. Sediments formed slowly, so immense stretches of time must have been required to create them. Hut-

ton had to let go of the chronologies of Ussher and others, relying instead on observations and logic. The Earth, he realized, must be at least many millions of years old, and it had experienced cycle upon cycle of sedimentation and upheaval.

Hutton's unconformity. This rock formation in Scotland provided evidence for Hutton's hypotheses about the age of the Earth and the processes by which different types of rocks had arisen. *(Farnham Geological Society)*

While many scientists agreed with his analyses of how rocks formed, and the idea that volcanic activity and underground heat could thrust huge masses of rock upward, the idea that this had happened over many repeating cycles contradicted the biblical account of history. Genesis spoke of one flood, not hundreds or thousands. Hutton needed better evidence to support his claims. During an excursion to a windswept region of the Scottish coast, he discovered a set of strata that seemed to tell the whole story. Several layers had formed on top of each other, had then been thrust upward by heat and pressure from below, and were finally covered by new layers.

Hutton published his ideas in 1795 in a huge, three-volume work called *An Investigation of the Principles of Knowledge and of the Progress of Reason.* His ideas might have had more impact, but the book had 2,139 pages and even his well-meaning colleagues found it hard to read. Mathematics professor John Playfair (1748–1819), who accompanied Hutton on some of his excursions, wrote: "The great size of the book, and the obscurity which may justly be objected to many parts of it, have probably prevented it from being received as it deserves." Playfair helped ensure Hutton's place in history by summarizing the geologist's ideas in a book of his own, called *Illustrations of the Huttonian Theory of the Earth,* published in 1802. One of the readers would be a young Scottish geologist of the next generation, Charles Lyell (1797–1875).

THE "NEW GEOLOGY": LYELL AND DARWIN

In the early 19th century, Hutton's hypothesis that the surface of the Earth had been reshaped by natural forces over long periods of time—possibly including the biblical flood—had spread among the scientific community. But *creationists* still had the upper hand on most of the field of geology. Lyell, who would later become one of Charles Darwin's closest friends and a strong supporter of evolution, would change that. He became convinced of Hutton's theories through careful studies of volcanoes in Scotland and France, excursions that gave him even more radical ideas.

Lyell proposed that a volcanic event could simultaneously raise masses of land and depress surrounding areas, perhaps as much as hundreds of feet. He found evidence of such events in Italy. Such upheavals could significantly alter a coastline, quickly making freshwater lakes of areas that had been ocean and vice versa. That, in turn, would explain why fossils of freshwater and ocean creatures could be found in alternating strata. Lyell presented his ideas on the formation of the Earth's surface in a series of books entitled *Principles of Geology,* published in the 1830s. Robert FitzRoy (1805–65), captain of the *Beagle,* gave

Charles Darwin the first volume as a present when the young naturalist joined the ship's crew. The second volume was published as the *Beagle* was en route and Darwin had it shipped to himself in South America. By that time Lyell was thinking even more widely, even speculating about how species might change over time.

His ideas had a strong influence on Darwin, even though the young man had been warned about Lyell by his geology professor. Christ's College in Cambridge, where Darwin enrolled after leaving Edinburgh, was still in the grip of traditionalists who wanted science to serve religion. Darwin took geology courses under professor Adam Sedgwick (1785–1873), who took him on field trips through Great Britain. Once Darwin found a fossil shell in a fairly recent layer of gravel—where it should not have been, according to Smith's rules of the connections between fossils and strata. He rushed it to Sedgwick, who was not impressed. The shell had gotten there by accident. "Nothing before had ever made me thoroughly realize, though I had read various scientific books, that science consists in grouping facts so that general laws or conclusions may be drawn from them," Darwin wrote. It was precisely what he would do in formulating the theory of evolution, many years later.

Sedgwick gave him an assignment: a 70-mile (112-km) walking tour in which he was to report on the arrangement of a particular stratum of rock in the northern part of Wales. Darwin returned with evidence that overturned a common hypothesis about the formation of that particular layer. The excitement of discovering new facts and fitting them into theory was a turning point in his life, again pushing Darwin toward geology. Later, when he was one of the first modern Europeans to witness an earthquake and tsunami firsthand, he had the necessary background to understand their importance for geological science.

THE PROBLEM OF FOSSILS

The 18th century and its great expeditions across the globe brought along a fever of collecting in England—the rich prided

Shrimp fossil from the Cretaceous period *(Mila Zinkova)*

themselves on gathering exotic birds, butterflies, plants, and perhaps most mysterious of all, fossils. Curiosity-seekers and scientists learned to iden-tify the strata where fossils were likely to be found, collected them, and sold them to collectors. Part of the attraction lay in the fact that no one knew what they were. They clearly re-sembled strange plants and animals that had somehow turned to stone—but how could that happen?

The Frenchman Georges Cuvier (1769–1832) had a brilliant insight into the nature of fossils by noticing the strong connec-tion between an animal's diet and its anatomy. The bodies of living carnivores were entirely designed to capture and devour animal prey, he wrote, and this gave them a structure entirely different than that of plant-eating animals. Fossil animals should obey the same principles, so Cuvier began an intensive study of the new finds. He became an expert of looking at single teeth or fragments of bones and drawing conclusions about their origins. He also discounted stories of mythical beasts, such as mermaids, by showing that they violated this principle of integrity. In 1817 he published a major book called *The Animal Kingdom,* in which he stated, "Every organized individual forms an entire system of its own, all the parts of which mutually correspond, and concur to produce a certain definite purpose."

Cuvier's ideas and his encyclopedic knowledge of anatomy throughout the animal kingdom won him wide international recognition. Yet he was focusing on just one part of a much larger picture, and this brought him into conflict with colleagues such as Jean-Baptiste Lamarck (1744–1829) and Étienne Geoffroy Saint-Hilaire (1772–1844). Geoffroy had helped Cuvier obtain a position at the National Museum of Natural History in Paris; the two became friends and wrote articles together. But over time Geoffroy had become interested in a different aspect of animal anatomy: the similarities between species. He discovered, for example, that fish have a structure similar to the wishbone of birds—why? Its functions in the bird—to support the stresses of flying—were unnecessary in fish. Geoffroy found an explanation: "It is known that nature works constantly with the same materials. She is ingenious to vary only the forms. . . . One sees her tend always to cause the same elements to reappear, in the same number, in the same circumstances, and with the same connections."

This principle led Geoffroy to some amazing insights, such as the fact that the bones that allowed fish to breathe were highly similar to the inner ear of mammals. Modern scientists recognize this as a fundamental component of how evolution works; nature builds on what is already there, and the great diversity of life stems from a much smaller number of structures that have been adapted in different organisms. Geoffroy took this much farther, proposing that all animals are built on a single mold, made different because the various parts have been stretched or rearranged. So far he was only talking about the *vertebrates*—animals with spines—but he was about to make a much bolder move and claim that the same principles held for

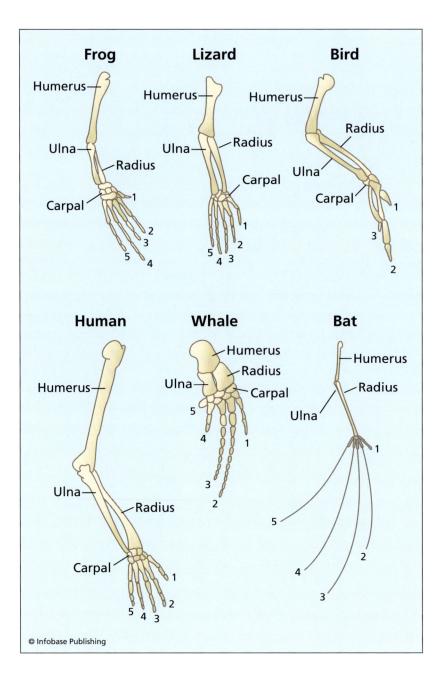

invertebrates such as mollusks and insects. They were so different, he understood, because insects lived "within their bones," whereas vertebrae lived "outside of them." It took the develop-

(opposite page) Careful anatomical studies of living animals and fossils at the beginning of the 19th century revealed unexpected homologies—parallel structures in different organisms that are usually arranged in the same way and perform similar functions. Before the theory of evolution there was no scientific way to understand such parallels. These diagrams reveal homologies between the limbs of tetrapods ranging from humans to bats to whales.

ment of molecular biology in the 20th century to prove that Geoffroy was right. This sounds like evolution, but Geoffroy had not understood that the forms represented a historical process—the ancient family relationships between animals. He was grouping facts but did not have a general principle by which to understand them.

Cuvier felt that Geoffroy Saint-Hilaire was going too far in drawing parallels between vertebrates and invertebrates. If animals had similar body plans, he felt, it was only because their bodies functioned in similar ways. The two men held a public debate on the issue. Most of those who attended believed that Cuvier won. But the debate was skirting the real issue. In a way both men were right and both were wrong, but there was no way to unify their positions without a theory of evolution.

Cuvier was equally interested in geology. In the regions around Paris he discovered traces of numerous cycles of rising and receding oceans, signs of ancient catastrophes that had occurred over very long periods of time. Marine fossils were sandwiched between layers containing the remains of land plants and animals, including monstrous creatures that no longer seemed to exist anywhere. This was obviously troubling to religious thinkers, who came up with ingenious theories: Fossils might somehow form themselves out of structures in the soil, the way that trees grow from seeds. Most scientists still believed in *spontaneous generation,* the idea that simple organisms such as worms or flies could arise by themselves out of rotting meat and other food. Until Pasteur disproved the idea, it did not seem so strange that animallike forms should somehow crystallize, all by themselves.

Reverend William Buckland (1784–1856), an avid fossil collector in Oxford and the first person to publish a scientific description of a dinosaur, waged a lifetime campaign to prove that the fossils in strata confirmed the order in which animals were created in the Bible; fish and lizards were created before mammals, for example. He believed that the creation story could be saved if biblical days were reinterpreted as much longer "creative epochs." When he became the first professor of geology at Oxford he gave a speech subtitled, "The Connexion between Religion and Geology Explained." It was a reaction against an increasing appreciation of the ideas of Hutton and Lyell about geological time and was well-received by conservative colleagues.

But Buckland could not explain alternating phases of sea- and ocean-living creatures, and no one knew what to make of monstrous dinosaurs. When he became director of Britain's largest museum, the Ashmolean, he invited Georges Cuvier to have a look at the fossils in the collection. Cuvier stated that at least one of the dinosaur specimens came from a land-living lizard that ate plants and was more than 40 feet long. Buckland could not fit this into his religious view of the world and waited many years to annouce the discovery of the first land-based dinosaur.

Buckland's situation was typical of most researchers of his time; many excellent scientists were desperately trying to reconcile new discoveries in their fields with the Bible's description of creation. But the tide was turning. Most serious geologists and paleontologists were moving away from a strict interpretation of the Bible, and the idea that animals could become extinct suggested that new types might also be born. Now that the Earth was thought to be very old, the stage was set for supposing that species might change over immense periods of time.

JEAN-BAPTISTE LAMARCK AND THE TRANSMUTATION OF SPECIES

Georges Cuvier's older colleague Jean-Baptiste-Pierre-Antoine de Monet, chevalier de Lamarck (1744–1829) is a curious figure of history who is remembered best for being wrong; his name

Erasmus Darwin and the "Living Filament"

Erasmus Darwin (1731–1802)—physician, inventor, poet, philosopher, writer, and grandfather of Charles—was an eccentric and brilliant doctor who studied the animal world and thought deeply about life. In the second half of the 18th century he started to ponder fossils, coming to the conclusion that living species had descended from simpler forms of life. Part of his inspiration came from the work of Georges-Louis Leclerc, comte de Buffon (1707–88).

Buffon had fled France to escape arrest after an impetuous love affair and a duel. In England he discovered Newton's methods of doing science and began thinking about life in a new way. When he finally returned to France, he established a research center and began working on a 44-volume encyclopedia called *Natural History* that would be published between 1749 and 1778 and cause a scandal. Not only did he promote the idea of an ancient Earth, but he also proposed that species had changed over time. The Catholic Church condemned him, calling him an atheist and burning his books.

While Cuvier had emphasized the features of animals that fit them to their lifestyles, Buffon focused on strange features that did not seem to have any logic at all. Animals were far from perfect, he wrote, citing many examples of poor design. Pigs, for example, had toes that did not touch the ground. This made no sense unless they had been inherited from earlier forms of animals in which the toes functioned. All mammals had similar limbs, in fact; could they have had a common ancestor? The Church was enraged by his work, but Buffon was not about to suffer the fate of Galileo; he wrote in a slippery style and built

(continues)

(continued)

in an "escape clause." Facts and observations might seem to lead to logical conclusions, he wrote, but only a person who had never read the Bible would believe them. Everyone else, of course, would understand the truth.

The debate between Cuvier and Buffon raised two issues that would have to be solved by any scientific theory of life: first, each type of animal shares features with many other species, and second, that each is adapted to its own environment. Erasmus Darwin tried to expand on Buffon's ideas. At the end of the 18th century he published a major book called the *Zoonomia* in which he wrote the following:

> In the great length of time since the Earth began to exist, perhaps millions of ages before the commencement of the history of mankind, would it be too bold to imagine that all warm-blooded animals have arisen from one living filament, which THE GREAT FIRST CAUSE [sic] endowed with animality, with the power of acquiring new parts, attended with new propensities, directed by irritations, sensations, volitions and associations; and thus possessing the faculty of continuing to improve by its own inherent activity, and of delivering down those improvements by generation to its posterity, world without end!

The beginning of the text sounds like an evolutionist's account of the descent of life from an ancient, very simple "filament." But the end of the passage shows that Erasmus Darwin, like other thinkers of his time, fundamentally misunderstood heredity. Many found Erasmus's philosophy attractive, but he could not back it up with facts, and besides, in the wake of the French Revolution, the society of Great Britain was wary of radical new ideas.

usually appears as a footnote to the story of Charles Darwin and evolution. Lamarck made the famous claim that giraffes got their long necks by stretching for the most luscious leaves, high in the treetops. He believed that each generation passed its slightly elongated neck to the next, finally achieving the towering height of today's animals. Lamarck's assumptions were wrong, but his theory that species could transform themselves was extremely influential.

Lamarck was influenced by the ideas of Buffon, whom he knew, and drew on Cuvier's ideas that species' bodies had functions connected to their environments and lifestyles. He also knew of Erasmus Darwin's work, and was influenced by the *"Great Chain of Being,"* a philosophy stretching back to the ancient Greek philosopher Aristotle. This system claimed that life formed a ladder from very primitive to extremely advanced organisms, with every species trying to move upward and improve itself. Higher and lower rungs represented greater and lesser levels of perfection. Man, with his ability to reason, held the top position, just below the gods. As organisms strived to move upward, they acquired new forms and behavior. Lamarck believed that they could somehow pass these characteristics along to their young. In doing so, he made two mistakes: He saw evolution as a process of improvement, and he believed that lifetime experiences and learning somehow entered an organism's hereditary information.

One might think that Cuvier and Lamarck, working in the same museum, might have found a way to unite their discoveries into a single, more powerful system. Instead, they became bitter rivals. Cuvier regarded Lamarck's ideas as ridiculous; animals' organs were obviously already "in tune" with each other and could not—or should not—change through use. There was no evidence, he claimed, that transformations of species had ever happened. (If he had talked to farmers and animal breeders, he might have thought differently.)

Lamarck's ideas addressed serious questions about the relationships between species while offering a way to keep humans at the pinnacle of creation, a step below God. It was a type of wishful thinking that was appealing to both progressive social

Lamarck's View of Species Change

Natural Selection as Defined by Darwin and Wallace

© Infobase Publishing

thinkers and those who wished to justify slavery. One could interpret it to mean that different human races occupied different
rungs on the ladder of perfection, where higher forms ought to

(opposite page) Differences between Lamarck's and Darwin's ideas of species change. Left: According to Lamarck, animals' needs and desires could change their behavior, which could produce inheritable changes in their bodies. Giraffes, for example, would need to stretch to reach high leaves when they had eaten all the low ones, causing their necks (and those of their offspring) to become longer. Right: Darwin's natural selection works differently. A population of giraffes will naturally have a few animals with longer necks. If only they can get food, they will survive to pass along genes for long necks to their offspring. Giraffes that are not tall enough die out. The entire species may become extinct if the population does not have much variety, and none of the animals can obtain food.

have dominion over lower ones. Darwin's theory would overturn this idea, but even today, some people misunderstand evolution by thinking it means that species improve themselves, for example by supposing that intelligence "had to evolve." This probably reflects a long history of confusion between Lamarck's ideas and those of Charles Darwin.

A simple experiment could have disproved Lamarck's concept of the inheritance of acquired traits, and several decades later, the German biologist August Weismann (1834–1914) performed it. He cut off the tails of mice for generation after generation. No matter how long he did so, none of his animals gave birth to a litter of mice lacking tails. The experiment has been repeated by many other scientists and has always led to the same conclusion.

NATURAL THEOLOGY AND "INTELLIGENT DESIGN"

At the beginning of the 19th century, most theologians and scientists still hoped that new discoveries could be reconciled with religion and the Bible, but it was becoming harder. A theologian named William Paley (1743–1805) hoped to bring science back on the right track, laying the groundwork for a religious movement that is known today as "intelligent design." *Natural Theology, or Evidences of the Existence and Attributes of the Deity*

Collected from the Appearances of Nature, the title of Paley's 1802 book, sums up its thesis: The aim of science, in Paley's mind, should be to prove the existence of God. Since God created the universe, whatever scientists discovered should strengthen the case for His existence. If science presented what appeared to be contradictions, it would be because science or human understanding had not yet progressed far enough to completely decipher God's design. Paley's book begins with the following famous watchmaker analogy:

> Suppose I had found a watch upon the ground, and it should be enquired how the watch happened to be in that place. . . . When we come to inspect the watch, we perceive . . . that its several parts are framed and put together for a purpose, e.g. that they are so formed and adjusted as to produce motion, and that motion so regulated as to point out the hour of the day . . . the inference, we think, is inevitable; that the watch must have had a maker; that there must have existed . . . an artificer or artificers who formed it for the purposes, which we find it actually to answer; who comprehended its construction, and designed its use.

The same thing, Paley wrote, held true for everything in the natural world. Pain existed so that people would take care of their bodies, and so that they would feel good when it stopped. Even night, he claimed, had been designed to an exact length— to fit humans' need for a certain amount of sleep!

Paley was in his way a very scientifically minded man, and his book contains careful observations of how things in nature fit each other, which he interpreted as evidence for the existence of God. Yet as attractive as Paley's arguments were to some, he had a great number of critics. There was no guarantee (except for his personal belief in God) that scientific discoveries would not one day contradict scriptures. And in a way, natural theology was planting the seeds of its own demise. It encouraged men of religion to pursue science; some of them eventually discovered things that undermined their faith.

The argument for intelligent design reminded some of the character Doctor Pangloss in Voltaire's novel *Candide*. Pangloss preaches that "this is the best of all possible worlds," and uses twisted logic to cope with anything that does not seem perfect. His pupil Candide, forced out of a comfortable home and into the harsh world, does everything he can to hold on to this notion, but finally he has to admit that things might not be so perfect after all. Something similar happened when scientists scrutinized nature. The design of human teeth, for example, left much to be desired—especially in an age before anesthetics and modern dentistry. To be fair, Paley was not necessarily claiming

Proponents of the religious philosophy "intelligent design" have used the pocketwatch as an analogy for living creatures for several hundred years. *(Empire Jewelers)*

that everything was good, or even that it was perfect—but that it was designed. That required a Designer.

Intelligent design also ran up against a straightforward, well-known fact: people already knew that extremely complex structures such as the eye could arise "by themselves" from much simpler origins. This happens in every embryo, which grows from a single cell. To get around this problem, Paley had to turn the watch in his analogy into a miraculous instrument which could somehow build and reproduce itself. It would have to contain an entire inner world of "lathes, files and other tools." The fact that such a watch had never existed did not matter, Paley wrote; at some point, there had to have been a powerful master Creator to make it, just as Descartes claimed that the eternal soul could only have stemmed from a more perfect being.

Natural theology is important because it shows how serious the conflict between religion and science was becoming. Science of the early 19th century was an easy target for Paley and his followers, because new hypotheses were still based on attractive ideas rather than facts. As long as there was no real scientific alternative to the Bible's account of the origins of life, natural theology could still hold its own. Its adherents hoped to turn the clock back and steer science away from Newton's methods, but they were reasonable enough to keep their eyes open and appreciate scientific discoveries.

The approach might have seemed scientific at the time, but it is completely different from how science is practiced today. Paley started from the premise of a God responsible for creating the universe, and in his mind, nothing would ever demolish this hypothesis. It made science into something like a baseball game whose outcome is decided before the first inning; the purpose of playing is to ensure that the right team wins. If things start to go wrong, umpires have to step in and change the rules. Today's scientists are not at liberty to bend facts any way they like; it is always possible for theories or hypotheses to be demolished by the results of experiments. That has happened over and over again, even regarding hypotheses within the field of evolution.

Natural Theology was so influential that it remained on the required readings list at many universities, even into the 20th century. It is a great irony that the man who would deal the strongest blow against Paley's life's work attended the same college where Paley's ideas ruled for so many years. It is an even greater irony that when Charles Darwin moved into the college in 1828, he was given the same rooms that William Paley had lived in a few decades before.

LINNAEUS AND THE CLASSIFICATION OF LIFE

The 18th century has been called the "Age of Classifiers" because so many people were consumed with a passion for categorizing things, particularly living creatures. In every mound of horse dung and under the bark of every tree, new species of beetles might be found, and searching for them became a sport. Half the joy was the search; the rest was classifying and naming the new creature.

The preeminent classifier of the Enlightenment period was the eccentric Swede Carolus Linnaeus (1707–78), a professor and pastor. Linnaeus created one of the first exacting methods for classifying plant species. Whereas previous classification systems focused on flowers and leaves, Linnaeus zoomed in on the reproductive organs of plants (reflecting a personal preoccupation with the relationship between the sexes). Given the important role that sex plays in evolution, this turned out to be a good decision.

Linnaeus's system classifies species into groups based on the number of male organs—the *stamens;* the number of female *pistils* is used to describe subgroups. Linnaeus extended plant metaphors to thinking about people and societies, even describing his relationship with his wife in botanic terms. Such double entendres remained a style in the science of botany for decades afterward, even though Linneaus was criticized for turning botany into a mixture of science and romantic poetry.

Linnaeus believed that species changed, not in the sense that new ones could arise, but that plants could become better attuned to their environments by drifting away from their original perfection. The Garden of Eden, he claimed, was a place of great harmony between all of its inhabitants. As plants, animals, and humans scattered across the globe, they somehow lost this harmony. It might be possible to bring it back, and Linnaeus tried to do this in experiments conducted in Sweden rather than Babylon. He proposed that his experiments could make the country into a new Eden—self-reliant, independent of foreign wares and trade—arguments which helped him win a great deal of money from the Swedish government.

He conducted thousands of experiments on plants sent back by collectors who had been dispersed across the globe, creating the beginning of a worldwide scientific network. Many of the specimens (and some of their collectors) died during the arduous voyages to Sweden; other plants perished because they had been brought to the wrong climate. Species, it turned out, were not as adaptable as Linnaeus thought—at least not with the methods he applied.

Linnaeus had a huge influence on biology. He inspired generations of young naturalists to collect specimens and attempt to classify the entire living world. People began to study the features of organisms carefully and scientifically, mapping them into a great system of life. At the time, they did not realize that by comparing and grouping things, they were also drawing a map of the family relationships between plants and animals. Linnaeus had a great influence on Darwin's grandfather Erasmus, an ardent "Linnaean" who wrote poems comparing the reproductive behavior of plants to a Tahitian marriage ceremony.

Another English naturalist of Erasmus's generation, Joseph Banks (1743–1820), was likewise a devoted follower of Linnaeus who discovered a great number of new plant and animal species while traveling with the explorer James Cook. On a scandalous trip to Tahiti, Banks had amorous adventures with the native women that made the headlines back home. Even so, upon returning to Great Britain, Banks was able to use his political connections to become appointed President of the Royal Society, England's most prestigious scientific group. He held on to that position for four decades and used it to increase government support for science. At the same time he developed an almost dictatorial control over opinions and theories. Banks's influence was felt for many years after his death—even into the early 1830s, as a ship called the *Beagle* was outfitted for a round-the-world voyage to further England's military, political, and scientific aims. Largely because of Banks's work, even military expeditions usually included a naturalist. Charles Darwin would be one of them.

THOMAS MALTHUS AND THE THREAT OF OVERPOPULATION

To judge from the number of eggs that animals lay, it is a wonder that the Earth is not knee-deep in sea slugs, or smothered in mile-high anthills. Most species, including humans, produce far more offspring than are necessary to maintain a stable population. This was common knowledge in biology, but until the beginning of the 19th century, there was no theory to explain the forces that kept species in check. Thomas Malthus (1766–1834) changed that when he considered human population growth in *An Essay on Population,* written in 1798.

Born with a cleft lip and palate, Malthus led a sheltered childhood and was schooled at home by friends of his father, a wealthy scholar who knew some of France's most prominent philosophers and scientists. As an adult, Malthus first took a job as a country curate and was then offered a position as professor of history and political economy at Jesus College in Cambridge, England, in 1793. Like the natural sciences, the social sciences of Malthus's era were dominated by religious ideas: The ultimate goal of economics and politics should be to create a perfect, moral society.

In Malthus's day society was moving in a direction that seemed simultaneously promising and threatening. The industrial revolution was sweeping Europe. Technology and industry ought to help bring people close to perfection, it was believed, but their effects on society were not all positive; there was a growing gap between the rich and the poor. Malthus was deeply aware of rising poverty and hoped that his studies of society would lead to a solution. Instead, what he learned gave him a pessimistic view of man's future. Human population, he observed, would always grow much faster than the production of food and other necessary resources. People reproduced in a "geometrical" pattern; every person who is born can have many more children. The food supply, on the other hand, grew in an "arithmetical" way, limited by realities such as the availability of land and labor and the productivity of plants. At some point, populations would outgrow their ability to feed themselves, and the consequence would inevitably be poverty, famine, wars, and disease. In a world without birth control, there was no obvious way to limit the birthrate except through "self-discipline," and Malthus did not hold out much hope of humans acquiring that. Even now in an age of convenient birth control, humanity has not curbed its growing population. Many modern thinkers echo Malthus's concerns. The huge public reaction to *The Population Bomb,* written by Paul Ehrlich in 1968, resembled how people responded to Malthus's essay.

Malthus's observations had a significant impact on politicians. Previously, an increasing population had been thought of only as a way to acquire more wealth, just as people might have more children so that they will be supported in their old age. Great Britain had passed "poverty laws" to attempt to adjust the standard of living between rich and poor, but some of the measures were actually encouraging poor families to have many more children. This was increasing the problem.

Malthus's essay was read several decades later by the young Charles Darwin and Alfred Wallace. One of their greatest accomplishments was to understand that the forces that keep human populations in check worked throughout nature and that their ultimate effect is to create new species. Both men said that Malthus's work provided an essential piece in solving the puzzle of the origin of new species.

The development of the scientific method and the thinkers of the 18th and early 19th centuries set the stage for evolution. Astronomers and physicists had discovered that the matter of the universe was governed by invisible forces that could be expressed as laws. Geologists claimed that the Earth and life were very old, and scientists had begun to hypothesize that species could change. What remained was to discover the hidden laws that, over huge periods of time, governed life—and possibly had created it in the first place.

2

Building the Foundations: The Voyages of Darwin and Wallace

From the 16th century onward, English ships traveled to the far reaches of the globe on expeditions of discovery and colonization. Each expansion increased the wealth of the British Empire through trade and the exploitation of natural resources. The expeditions helped advance British science by bringing back exotic new plants, animals, and fossils for study. At the beginning of the 19th century, scientists often accompanied naval voyages in hopes of finding plants that could be sources of new foods and textiles or other valuable resources. Robert FitzRoy, captain of the British navy ship *Beagle,* wrote the following in his journal of an expedition to South America:

> There may be metal in many of the Fuegian mountains, and I much regret that no person in the vessel was skilled in mineralogy, or at all acquainted with geology. It is a pity that so good an opportunity of ascertaining the nature of the rocks and earths of these regions should have been almost lost. . . .

The next time he sailed, he took along a young geologist named Charles Darwin.

The fascination with nature also offered opportunities to young men like Alfred Wallace, willing to travel the globe in search of plants and animals for wealthy collectors and museums. Such voyages were full of danger and hardships. Wallace suffered from recurrent fevers; Darwin nearly died of dehydration, freezing, drowning, and disease. Throughout most of his life he suffered from a strange illness that may have been caused by a tropical parasite. Darwin's friend Joseph Hooker, who traipsed through the Himalayas for years on plant-seeking expeditions, was thrown into jail after crossing the border into Tibet in search of new species of rhododendrons.

These voyages were essential to developing the theory of evolution, partly because they were opportunities to observe species in habitats that had not yet been significantly altered by human activity. A great mystery surrounded the way plants and animals were spread across the globe; it was hard to understand why God had not simply put the same species everywhere. Explorations also provided the chance to observe humans in other types of relationships with nature, which might answer questions about the history and development of society. People were curious about the connection between technology and culture and whether certain forms of government were better than others. Would a perfect society come through advances in science, or should people try to return to a "state of nature" like the Garden of Eden? Conquerors and travelers found very few examples of these romantic ideals of paradise. At the same time, their own behavior made it hard to believe that technological progress would lead to a more moral society.

THE VOYAGE OF THE *BEAGLE*

The *Beagle* was built in about 1820 and was one of several ships sent to South America during the late 1820s on surveying missions. The navy needed accurate maps of coastlines and harbors to protect British ports abroad and to navigate into new areas

HMS *Beagle* at Tierra del Fuego (Conrad Martens) from *The Illustrated Origin of Species* by Charles Darwin, abridged and illustrated by Richard Leakey *(Wikipedia)*

for trade. This was a huge task that would take years to accomplish. Lieutenant Robert FitzRoy was responsible for much of the mapping activity on the *Beagle*.

Two events during the *Beagle*'s original mission had a huge impact on her next voyage. The *Beagle*'s first captain, Pringle Stokes, committed suicide. A captain's job was hard and mostly administrative, a far cry from the image of a romantic figure standing on a windswept prow, charging into battle. He spent most of his time filling out paperwork and meting out punishments. To maintain authority he had to remain aloof from the crew, meaning years of loneliness. After the suicide, FitzRoy was given command of the ship.

The second incident involved the theft of a boat by natives of the Tierra del Fuego, the land at the southernmost tip of South America. While trying to recover the boat, FitzRoy and his men took two captives that they pressed into service as translators. FitzRoy decided to take the prisoners and two more Fuegians to England, where they could be educated. He promised to bring them home on the next surveying trip. He had in mind a social experiment to see if "savages" could be civilized. If so, what they learned might help improve their society when they returned home. Once the ship returned to Great Britain, it looked as if FitzRoy would be unable to keep his promise, because plans for a follow-up voyage had been cancelled. In the end, an influential relative pulled strings with the Admiralty, and the survey was on the agenda again.

FitzRoy was a precise and scientific man and had the *Beagle* outfitted with the best possible equipment. He brought on board 22 elaborate timepieces, essential to making accurate maps and charts because they were the only method to measure the ship's longitude. The instrumentation was considered extravagant, so FitzRoy had to pay for much of it out of his own pocket. He even hoped to substitute the ship's iron cannons for brass guns, which would not interfere with the compasses, but his superiors would not agree. In the end, he bought the guns himself, smuggled them onto the ship, and threw the iron cannons overboard.

He also engaged extra personnel, including a technician to take care of the clocks and an artist to capture images of the voyage. Finally, he began looking for a scientist—preferably an educated person with whom he could have dinner. Members of FitzRoy's own family had been affected by madness and suicide, and he was concerned for his own mental health. An intelligent companion of his own social class on board might help him avoid the loneliness and fate of the *Beagle*'s former captain.

Darwin was only 22 years old when he received the invitation to accompany the *Beagle*'s surveying expedition, chiefly as a geologist. No one suspected that he was about to embark on one of the greatest adventures of scientific history. In many ways, he was an unlikely candidate to be offered the position and to achieve what he did. So far his studies had not been very successful; enrolled in medicine in Edinburgh, he found the normal courses tedious and dropped out. The invitation to join the *Beagle* expedition came in the middle of his studies of theology in Cambridge, where he was not doing much better.

His father, Robert, was a wealthy doctor and financier who had been paying for his son's studies at the university in hopes that Charles would finish a degree and settle into the comfortable life of a country pastor. It was a life that appealed to Charles, who was religious enough and knew that a pastor's duties could be easily combined with an interest in geology and nature. Robert hesitated to allow his son to go on the voyage, fearing he would never finish his studies. Charles wanted to go, although he admitted that he did not have the best qualifications to become a world-roving naturalist.

Yet some qualities made him exactly the right person at the right time. He had absorbed a lot of science while at Edinburgh through extracurricular activities. His time at the microscope and on geological excursions revealed a talent at research, self-discipline, and a seriousness that had been lacking in his studies. And he understood that science involved looking beyond appearances to understand the principles that lay underneath. Professors and friends strongly urged him to go on the voyage. Their combined support changed his father's mind, and the path was cleared for Darwin to change history.

Five years later, Darwin returned home far more mature, with a remarkable set of notes and specimens that would give him a solid footing for a career in science. The book that he published about his travels in 1839, *The Voyage of the Beagle,* is an abstract of much more detailed journals that he kept, recording observations and impressions of geological formations and life. At the same time the book is an adventure story with tales of wild horseback rides across plains and mountains, battles between colonial Spanish forces and natives in South America, and narrow escapes as Darwin nearly froze on mountaintops or died of thirst.

The book became an immediate best seller and is still required reading at universities throughout the world, though not directly because of evolution, which Darwin had not yet figured out at the time of publication. Instead, *The Voyage of the Beagle* is poised at a crucial turning point of science; its value lies in the snapshot it gives of how scientists understood the world just before evolution changed everything. To read it is like rereading a favorite mystery story and understanding all the clues long before the detective does. Darwin grasped much of what he

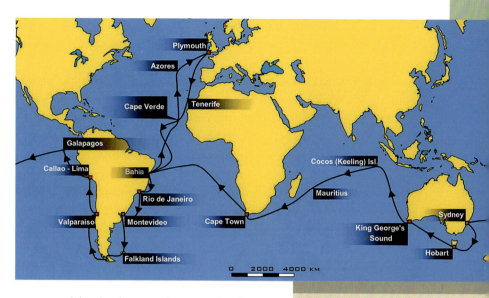

saw, and he had a good sense for key pieces of evidence, but he did not yet know how to assemble the elements into a complete scientific theory. His ideas were still dominated by the influ-

The route taken by the HMS *Beagle* from 1831–1836 (Wikipedia)

ence of natural theology and the cataloging spirit of Linnaeus. Yet a few years later at home, sorting through his collections, he would have answers to many of the questions that he posed on board.

Darwin's first encounter with the *Beagle*'s captain went reasonably well. FitzRoy was only 26, but he had a commanding, all-knowing presence that fit well with his absolute rule of the ship. There were only a few nagging issues. The navy man held to a common theory of the time that said personality could be predicted by the shape of a person's nose and the bumps on his skull. "He doubted whether anyone with my nose could possess sufficient energy and determination for the voyage," Darwin wrote in his autobiography. "But I think he was afterwards well-satisfied that my nose had spoken falsely." For his part, Darwin was shocked at the tiny size of the cabin he would inhabit. He would have to sleep in a hammock above his working table, which put him barely two feet from the ceiling. Then during a three-day practice run aboard ship, he was violently seasick. But nothing would make him turn back now. He was equally thrilled and terrified when, after numerous delays, the *Beagle* set sail on December 27, 1831.

FitzRoy and Darwin managed to get along well for five strenuous years, which contributed a great deal to Darwin's success. It was not always easy; Fitzroy could fly into a rage one day on an issue such as slavery (which Darwin opposed but the captain originally supported) and apologize the next. Darwin learned to make himself scarce when necessary, and for his part, FitzRoy often went out of his way to put Darwin ashore in interesting places. The two men grew fond of each other and after the voyage exchanged friendly letters—FitzRoy called Darwin "philos," for "philosopher," and teased him with details of his wedding plans; Darwin told stories of the wild welcoming parties on his arrival home.

GEOLOGICAL DISCOVERIES, FOSSILS, AND EXTINCTIONS

During his five years on the *Beagle,* Darwin investigated everything: geological formations, the physical characteristics of plants and animals, the origins of coral, the social behavior of insects and other creatures, the backs of turtles, and the cultural clashes of South American natives and Europeans. One of the *Beagle*'s first stops was the island of St. Jago, 300 miles (480 km) off the coast of Africa. While strolling the beach, Darwin discovered a white layer of sediment about 30 feet (9 m) above the waterline. It was full of fossil oyster shells nearly identical to their present-day relatives, strewn along the beach below. He wondered about the age of the stratum and why it lay so high above the sea line. There were only two possibilities: The level of the ocean had dropped, or a force had pushed the land upward.

With this first landfall he had stepped into the middle of one of the major controversies of his time. Sedgwick, Darwin's geology professor in Cambridge, supported a *"catastrophe theory"* of geology that tried to align modern findings with the bibli-

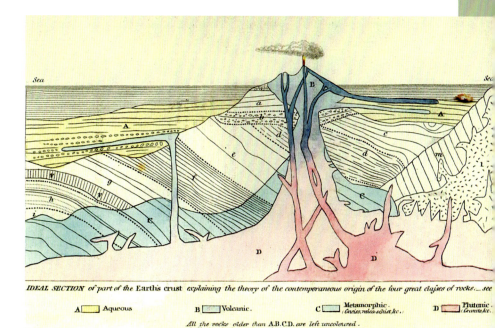

IDEAL SECTION of part of the Earth's crust explaining the theory of the contemporaneous origin of the four great classes of rocks.— see

A ☐ Aqueous.   B ☐ Volcanic.   C ☐ Metamorphic. *(Gneiss, mica-schist,&c.)*   D ☐ Plutonic. *(Granite,&c.)*

All the rocks older than A.B.C.D. are left uncoloured.

cal account of creation. It accepted that the Earth had been through major upheavals, but this had happened in a distant past, the days of creation and the great flood, a special time when unusual forces were at work. Sedgwick had warned Darwin away from the opposing camp, *"gradualism,"* promoted by Charles Lyell in the first volume of his *Principles of Geology.* Lyell maintained that the Earth behaved according to laws which had operated over vast stretches of time and that the forces at work in the present were basically the same that operated in the distant past. This was a simple extension of Newton's "rules of reasoning" for science. Researchers often have to make assumptions about things outside their direct experience. In such cases, a scientist should assume that things he cannot observe behave as the things he can observe do. Without clear evidence, one should not suppose that there has been a change in the laws of nature, that miracles happen, or that distant objects behave differently than those close at hand. Gradualism was an application of these

The frontispiece from Charles Lyell's *Principles of Geology* (second American edition, 1857), showing the origins of different rock types *(Public domain)*

principles to geology, and evolution would apply them to the living world.

What Darwin saw on St. Jago supported Lyell's theories. The similarity of fossil oysters and the live creatures on the beach meant that the white layer was relatively recent. The layer was higher in some places and lower in others, but there were no signs of major cataclysmic events. Later Darwin was lying on the floor of a forest in Chile when he felt a massive earthquake, one which rocked the entire coast. The *Beagle* was in the harbor and quickly set sail for the major city of Concepción, 200 miles (360 km) to the north, to find the town and its cathedral in ruins. A massive tsunami had struck the coastline, killing thousands, tossing huge stones onto the shore, lifting great blocks of earth. All of these effects were obviously related, and Darwin tracked their source to volcanic activity off the coast. Comparing the new landscape to the old proved that the Earth could still experience cataclysmic forces; they were merely slumbering and could return at any time.

Almost immediately after the *Beagle*'s first stop in South America, Darwin began finding amazing fossils which raised a number of questions. Each continent contained a mixture of unique and shared species, but in some geological periods, certain animals did not appear at all. Other layers held bizarre forms of life that no human had ever seen and had never been mentioned in the Bible. One dinosaur discovered by Darwin would have been longer than the ship itself and had to be left behind.

Sometimes what he found in layers of rock bore almost no relation to present-day animals; other fossils were recognizable but strange. He sent home the bones of a rodent the size of a rhinoceros, anteaters as big as horses, and enormous cats. "It is impossible to reflect on the changed state of the American continent without the deepest astonishment," he wrote. "Formerly it must have swarmed with great monsters: now we find mere pygmies."

There was no clear reason why species should have disappeared. The Americas could obviously support vast numbers of animals. In just a few centuries, horses brought by the Spanish had produced hundreds of thousands of descendants. Darwin suspected that extinctions might have something to do with the environment. Although he was not yet ready to follow these ideas to their logical conclusions, he was thinking along the right lines, raising issues like the tendency of species to produce too many offspring. "Every animal in a state of nature regularly breeds," Darwin wrote, "yet in a species long established, any *great* increase in numbers is obviously impossible, and must be checked by some means." Discovering the means was the crucial missing step to arriving at the theory of evolution. A crucial moment arrived much later in the voyage when he had the chance to observe a place that had remained almost completely undisturbed by humans—the Galápagos Islands.

LIFE IN SPACE AND TIME

Through his observations of fossils and geology, Darwin realized that the animal populations of most regions changed significantly over time. He did not yet know that species themselves changed. Like many others, he regarded his grandfather's and Lamarck's ideas as interesting but unscientific. Many years later, in the preface to the first edition of *On the Origin of Species,* Darwin praised Lamarck for being one of the first modern scientists to suspect "that species, including man, are descended from other species. He first did the eminent service of arousing attention to the probability of all change in the organic, as well as in the inorganic world, being the result of law, and not of miraculous interposition." On the other hand, Lamarck made serious mistakes: "He likewise believed in a law of progressive development; and as all the forms of life thus tend to progress, in order to account for the existence at the present day of simple productions, he maintains that such forms are now spontaneously generated." In other words, Lamarck made the mistake of believing that species changed by improving themselves. *Bacteria* and very simple organisms still existed because nature produced new ones from scratch.

Darwin was more influenced by the thinking of Lyell, whose books spoke of the relationship between species and their environments but had not proposed anything like evolution. Lyell strongly disagreed with Lamarck's views that one species could somehow transform itself into another or improve itself. The fossil record was incomplete and did not prove that life had become more complex over time, or that new species appeared. Lyell did believe that species could undergo a certain amount of adaptation to the places in which they lived. If they failed to do so, he pointed out, they might die off.

Everywhere the *Beagle* sailed, Darwin witnessed how completely species were dependent on each other and their environments. In the Straits of Magellan he pulled up strands of kelp—seaweed—which grew hundreds of feet from the ocean floor to the surface. An incredible number of species depended on this plant. In the roots alone he found small fish, shells, cuttlefish, crabs, sea eggs, starfish, and dozens of other species. Other parts of the plant hosted mollusks, crustaceans, and other animals. From region to region of the ocean the species changed, but kelp played an equally important role everywhere. Darwin described it as a "great aquatic forest." He wrote, "Amidst the leaves of this plant numerous species of fish live, which nowhere else could find food or shelter; with their destruction the many cormorants and other fishing birds, the otters, seals, and porpoises, would soon perish also." Even human populations might be affected.

On land, species were just as dependent on each other. But they were oddly distributed around the globe, and Darwin could think of no reason why a Creator would have given England a species of snail that plagued gardeners by feasting on their salads, while a nearly identical strain in South America had no taste for salad at all. Lyell suggested that each species had been created in a single place, and then spread as far as it could until running up against some natural boundary. The idea of natural barriers would also become important in evolution.

As the *Beagle* sailed up the western coast of South America, along the long western shore of Chile with its impressive chain of mountains, Darwin was about to discover signs of the deep underlying logic he had been looking for. Intent on finding more evidence of the geological forces that had raised the mountains, he climbed into the Andes, ascending the western slope and coming down on the other side. The species on the east slope were quite different than those on the west. But they were remarkably similar to plants and animals he had seen much farther to the east, on the central plains of the continent.

Darwin suddenly realized that the high Andes formed a nearly total barrier between two zones of life. The western side was a small strip cramped between the mountains and the oceans; the eastern slope descended into plains that went on for thousands of miles. Absolute distances did not matter as much as barriers. Species could not cross the Andes, so they did not mix with each other. Those on the east had lots of room to spread. Simple geography, Darwin realized, could dictate which species inhabited a region. From the creationist point of view, there was no reason why this should be so. Everything could have lived everywhere.

Such thoughts occupied him intensely as the *Beagle* sailed onward toward the small group of islands called the Galápagos, a volcanic cluster of rocks several hundred miles off the west coast of South America. Other than a few traders and a small prison colony, the islands had never had any human habitants. Ships had been using them as a stopover point to take on water and huge tortoises that could feed dozens of men and could be kept alive for months in a ship's lower deck. Some of the islands were little more than craggy bits of volcanic rock; others sported lush tropical vegetation. But the unusual location between the cold currents of the south and the warmer waters of the north made the Galápagos home to a rich variety of other forms of life, from penguins and seals to flamingoes and tropical birds. Monstrous lizards and tortoises lived side by side.

Darwin was fascinated by several species of ugly iguanas with bizarre habits. Most of them were vegetarian. They lived on land but some dove into the sea for food. Darwin wondered if they swam between islands, some of which were separated by distances of 50 miles (80 km). He threw one iguana into the water over and over again, curious to see how comfortable it

Galápagos iguanas *(Wikipedia)*

was in the ocean (they generally dis-liked the water, diving in only when they had to eat). To find out about their diets, he brought animals on board the ship for dissections. This smelled so horrible that his shipmates forced him to move his operating room onto land.

The Galápagos were full of species of birds, lizards, and turtles so similar to each other that without careful study it was hard to tell them apart. The only difference between some of the 13 species of finches, for example, was the size and shapes of their beaks. Darwin regarded these things as simply odd until Nicholas Lawson, an Englishman watching over a small Ecuadorian prison colony, claimed that each of the Galápagos Islands had a unique collection of species with its own peculiar characteristics. Lawson could tell where any tortoise had come from based on the form of its shell—on one island tortoises had saddle-shaped shells, on another they were rounder and blacker. The evidence could be found in Lawson's own garden, where he used shells as flowerpots. This chance remark would lead to a dramatic change in Darwin's think-ing and it saved his collections; he had already begun to mix up specimens from several locations. It had never occurred to him that islands only about 50 or 60 miles (from 80 to 96 km) apart, within sight of each other, with similar climates and made of exactly the same types of rocks, might hold different inhabitants.

Darwin frantically began sorting out his collections and looking at the islands in a completely new way. The rest of his days on the Galápagos became a race to discover whether the pattern for the tortoises was also true for finches, iguanas, other animals, and plants. Later, he made charts and lists of species found in each location. Why did James Island have 38 species of plants found only in the Galápagos, and why did 33 of these species only exist on James Island? Why would each island con-tain so many wholly original species, even though it was so similar in climate and geology to the other islands? There had to be an extremely active "creative force" at work, he wrote, but he had no idea what it might be.

The Galápagos archipelago has been called "evolution's laboratory" because its special geographical situation simplified the problem of the development of new species to just a few variables. Similar processes were at work in the South Ameri-can rain forests, but there the ecosphere was far too complex to grasp. The Galápagos were far from the mainland, so they were rarely influenced by the arrival of new species from the outside. At the same time, they were just far enough from each other to isolate some of their populations, which is a key factor in evolution.

When the *Beagle* finally left the islands, its voyage had been under way for over four years. Darwin was becoming over-whelmed, but the pace of work did not let up. He would have the rest of his life to think; now it was necessary to collect as many facts and specimens as possible. He did not want to fall into the same trap as his grandfather, or Lamarck—building a beautiful theory that stood on philosophy, rather than facts. (His contemporaries later estimated that he possessed a hun-dred times as many facts as Lamarck had.)

His book on the *Beagle*'s voyage abandons the huge questions raised by the Galápagos very abruptly—practically in mid-sentence. But a seed had been planted in his mind: What was true in the Galápagos held for South America. A mountain range or a stretch of water could form a barrier to the spread of life. Very fine differences between species like finches—sometimes amounting to only a single feature—might be the rule rather than the exception. Darwin did not yet see the connection between small differences within a species and small differences between species, but he had begun to accept that organisms could become transformed and that the environment played an important role in this process.

Though he was eager to return to England, there were more stops to make—Australia, Tahiti, and the Cocos Islands. Darwin carefully observed reefs of coral so deep and massive that the ship could not locate the bottom on which they stood. Lyell had proposed that coral—huge colonies composed of billions of tiny creatures—stood on the rims of ancient volcanoes. Before ever seeing a reef, Darwin developed a different theory about their formation. The living parts of coral colonies always lay at a shallow depth in the ocean. As the land dropped or the ocean level rose, new animals built upward toward the surface, on the stony remains of their ancestors. If the opposite happened and the land rose, the height of the colony fell and they built outward. Thus a reef preserved a record of its own history: geological events and life combined to create living islands, ecospheres just as complex as kelp forests. Tiny organisms linked themselves in huge colonies to create vast, mountainlike islands that became habitats for thousands of other species. Everything was interdependent.

The voyage of the *Beagle* ended in 1836, and Darwin returned home. A huge amount of work faced him and a group of

(opposite page) Darwin was the first to understand that coral reefs are "living islands" built by organisms in response to changes in the environment. Most corals live in shallow waters, at depths less than 200 feet (60 m). When water rises (or the land sinks), coral grow upward; if the water level remains constant or sinks, the reef grows outward.

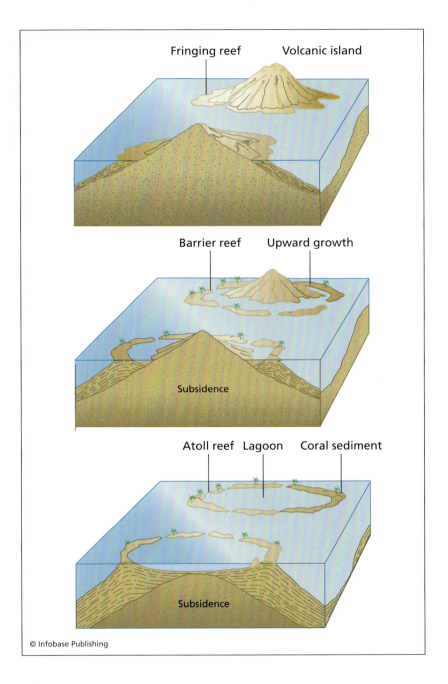

Fringing reef Volcanic island

Barrier reef Upward growth

Subsidence

Atoll reef Lagoon Coral sediment

Subsidence

© Infobase Publishing

experts that he had carefully selected to help; he had to preserve, catalog, and study thousands of specimens he had been sending home throughout the voyage. The ordeals of scrambling up mountains to collect species were over, and an entirely new type of ordeal was about to begin.

BARNACLES AND EXPERIMENTS

By 1844, Charles Darwin had understood the basic principles of evolution and written a brief sketch of the theory. Then he locked the draft in a drawer in his study, composed a letter to his wife explaining what to do with it in case he died unexpectedly, and devoted himself to one of nature's oddest creatures: the barnacle, best known for the problems it causes when it glues itself to the hulls of ships.

He anticipated that dissecting, analyzing, and classifying the many different species of this creature would take months, perhaps a year. But as colleagues all over the world sent him specimens, he entered a bewildering labyrinth of classification that cost him nearly a decade of his life. The exhausting work was made more difficult by the appearance of serious health problems. His doctor ordered enforced rest and sent him off on cures where he was wrapped, mummylike, in wet sheets for hours on end.

The whole time the theory lay locked away, never even mentioned except for hints in cautious discussions with a small circle of friends, some of whom became frustrated with his hesitation in publishing it. Darwin knew he needed a reputation as the world expert on one topic to have any hope of his theory being taken seriously. He chose the barnacle because of the immense variety of specimens found across the world—including some fascinating examples he had personally collected during his voyage on the *Beagle*. The creatures were poorly understood; no one had ever thoroughly classified them. And a careful study might make or break his ideas on evolution.

He was amazed and confounded by the creatures' strangeness. Upon prying open the shell of a female, he discovered that the shell's interior was divided into compartments. These served as apartments for her tiny mates, male barnacles that would live their entire lives in the female body, never seeing

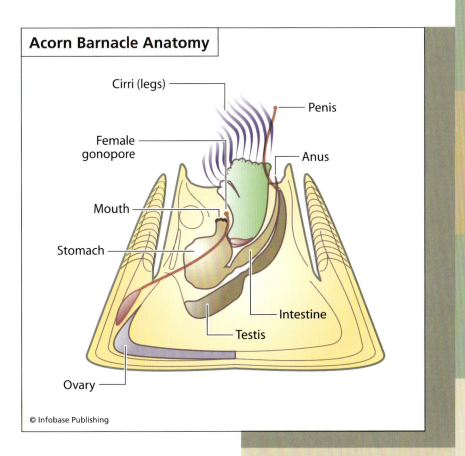

Acorn Barnacle Anatomy

Cirri (legs)

Penis

Female gonopore

Anus

Mouth

Stomach

Intestine

Testis

Ovary

© Infobase Publishing

the world outside. They were much simpler creatures than the female—Darwin called them "bags of sperm"—and their only function was to perpetuate the species. What an incredible solution to the problem of finding a mate, he thought. The males were not alone; they were surrounded by newborn larvae and barnacles in other stages of development. It was like peering into a miniature ocean, a new world inside a shell.

Studying the incredible variety of barnacles found throughout the world took Darwin eight years. The diversity of these strange animals gave him a close look at how natural selection works on varieties to create new species, adapted to a wide range of environments.

The work on barnacles plunged Darwin deep into one of the most crucial aspects of evolutionary theory. Either what he found would confirm his ideas or somehow contradict them.

The books he wrote about barnacles were two things at the same time: traditional works on anatomy and classification, and a reference for the future, a mass of evidence that would support evolution. There were thousands of types of barnacles alone. Which were distinct species, and how many were simply variants of the same type? Every type of barnacle he studied represented a new solution to the problem of how an organism could survive and reproduce within the possibilities offered by its environment.

Because each member of a species was slightly different, it was helpful to propose an *archetype*—a standard representative by which to measure all the others. He realized that this was really another question in disguise: He was really searching for the original type. Somewhere among all his specimens, he thought, might be one close to the ancestor of all the others, the first barnacle that had ever evolved. But how would a person know if he found it? His friend Joseph Hooker, climbing through the Himalayas, was having the same problems understanding varieties of rhododendrons and other plant specimens. Hooker knew that Darwin was getting close to a solution. By straightening out the family tree of barnacles, Darwin would be showing the way through a labyrinth of problems that other scientists would face time and time again with other organisms.

Nothing Darwin discovered in his work with barnacles contradicted his ideas about variation and natural selection. But a single species, even exhaustively studied, could not prove the theory, and it did not solve all the questions raised by evolution. For example, if all varieties of a plant descended from a single ancestor, there had to have been some way for them to spread throughout the world. Oceans and mountains might be too great a barrier.

Seeds might be carried by winds, in dirt clinging to the feet of birds, or carried along in their digestive tracts. Darwin needed to know if seeds could survive long voyages under these circumstances. He built aquariums in which he immersed seeds for weeks, retrieving them and then planting them; he did the same thing after extracting them from the droppings of birds and other animals. He found that they were able to withstand a wide variety of conditions. The experiments proved that life was amazingly resilient. Once it came to exist, he had no doubt it could find its way around the globe.

He also carried out intense studies of animal breeding and agriculture. If anyone ought to know about changes in species, he reasoned, it would be people who had spent their lives trying to improve crops or change the characteristics of horses, cows, sheep, dogs, and many other animals. A large section of his major work on evolution, *On the Origin of Species,* is devoted to breeding.

The weakest part of the theory was its explanation of how evolutionary changes could be transmitted from parent to child. Darwin carried out experiments on a wide variety of animals and plants, including over 40 species of peas, not knowing that at the same time a monk named Gregor Mendel was attacking the same problem with the same plant. They took very different approaches to the problem, with the result that Mendel discovered the laws that governed heredity, while Darwin learned almost nothing.

ALFRED RUSSEL WALLACE IN THE SPICE ISLANDS

Alfred Russel Wallace (1823–1913) was 14 years younger than Darwin and came from a much different background. His father was a gentleman with a modest regular income, but Thomas Wallace had none of the ambitions, sense of business, or family wealth that had made Charles Darwin's father into a rich man. A series of bad investments brought the family into severe financial difficulties. Alfred and his brothers constantly had to work to supplement the family's income and support their own education. His father died when Wallace was 20; by that time the son had tried watchmaking, surveying, and teaching school. Along the way he studied and read a great deal, including the works on geology that had so strongly influenced Darwin, and developed a passion for collecting plants. He also met Henry Bates, a young naturalist his own age. The two began collecting beetles and butterflies together.

Alfred Russel Wallace, codiscoverer of the theory of evolution (*National Portrait Library*)

After his father's death, Alfred made a living as a surveyor and moved to the town of Neath in Wales, where he became curator of a small museum and gave lectures on science. He continued his excursions into the countryside on surveying expeditions with one of his brothers and exchanged letters with Henry Bates, comparing the species they had found.

In 1845 he read a popular new book called the *Vestiges of the Natural History of Creation,* whose claims were so extravagant and controversial that the author (Robert Chambers) published it anonymously. The work began by suggesting if the entire universe was ruled by natural laws like gravity, then species must be ruled by such laws, too. He included observations from Geoffroy Saint-Hilaire and Cuvier and attempted to wrap them all up in a theory of life. For example, Chambers wrote, "While the external forms of all these various animals are so different, it is very remarkable that the whole are, after all, variations of a fundamental plan, which can be traced as a basis throughout the whole, the variations being merely modifications of that plan to suit the particular conditions in which each particular animal has been designed to live."

The book suffered from the same weaknesses as previous theories, offering neither any real evidence nor a mechanism by which animals adapted to their environments. Chambers agreed with Erasmus Darwin and Lamarck that changes in species represented improvements. He went farther, claiming that humans were a part of the scheme; they, too, had arisen from "less perfect" forms of life. He saw no difficulty in reconciling this idea with religion: "if He, as appears, has chosen to employ inferior organisms as a generative medium for the production of higher ones, even including ourselves, what right have we, his humble creatures, to find fault?"

Because of the lack of evidence, Darwin and most scientists rejected the book's speculations. Wallace was more open-minded. Either the author was right or wrong, he wrote to Bates; what scientists ought to do was look for evidence either way. The way to do it might be to select one family of organisms and spend years studying it very carefully, hoping to arrive at a new theory of the origin of species. This was exactly what Darwin was doing with his barnacles.

There was money to be made overseas searching for new species and sending back specimens to museums and private collections. Inspired by Darwin's experiences aboard the *Beagle* and the adventures of other travelers, Wallace and Bates decided to head for the Amazon. After several months of preparation, the two men set off on a merchant ship for South America. They were determined and successful collectors. In the field,

Wallace possessed an almost infinite patience, once spending nearly a month in pursuit of a single beautiful species of butterfly. The men were able to sell enough specimens to collectors back home to make a living. After several months, for reasons that they never revealed, they split up. Wallace found a new partner in his younger brother Herbert, who sailed across the Atlantic to join him.

In spite of several near disasters—including an infection caused by an accidental gunshot wound and a near-fatal case of malaria—Wallace's first expedition abroad was a success. While most collectors were regarded as the "legmen" of museums and rich curiosity seekers, Wallace was seen as someone who also thought deeply about the central problems of science, and the respect he earned improved the reputation of collectors in general. The voyage had a heartbreaking ending because the ship that was to take him and the majority of his collections back to England burned and sank. Nearly everything was lost, including Wallace's journal and notes. It was a crushing blow, and he vowed not to travel abroad again. (He would not keep the promise for long.)

Back in England he was able to reconstruct enough of his notes to write two small books, one about species of palms and their uses among the natives of South America, and a second account of his travels. During this time he briefly met Darwin, but since the older scientist was not talking about his theories yet, and the younger man had not formulated any, they did not make a very strong impression on each other.

Like Darwin, Wallace had almost universal interests. Although his books did not match the literary quality of the *Voyage of the Beagle,* and scientists criticized him for a lack of rigor—partly due to losing most of his notes—he gained enough of a reputation to finance another trip. He decided to head for the Spice Islands, a labyrinth that stretches between Indonesia and Malaysia, which had never been thoroughly explored.

He established a base in western Borneo, which turned out to be a biological gold mine. During one two-week period he found an average of 24 new species of beetles every day. On

While traveling in the Spice Islands, Wallace discovered a natural barrier now known as the Wallace Line. Different species are found on either side. Even many birds, which could fly across, remain on one side of the line. Some of the islands separated by it are only about 35 miles (56 km) apart, but the water is very deep. The division between species on either side probably began when the water was more shallow and many of the islands (except those separated by the line) were connected by land.

a single day he found 34 entirely new species. He caught his first sight of an orangutan, whose eerie, manlike eyes and intelligence fascinated Europeans. The hard work of collecting was occasionally interrupted by sickness or rains, and whenever this happened, Wallace sat in his hut to think and write. When the rains stopped or the fever subsided, he jumped into his boots and waded back into the forest, leaving his theoretical work until the next period of enforced immobility.

The incredible variety of highly similar species struck Wallace just as it had Darwin in the Galápagos, and he began making notes for a book to be called *The Organic Law of Change*—just as the Earth had to obey laws, so did species. In 1855 he wrote a paper called "On the Law Which Has Regulated the Introduction of Species," published in the *Annals and Magazine of Natural History.* In it he stated that species arose through gradual change. The evidence for this, he said, was that "every species has come into existence coincident both in time and space with a pre-existing closely allied species." Whatever the rules were by which species arose and changed, they were closely linked to space and time. Darwin read the paper but did not give it much attention. Wallace was not really saying anything that had not been stated before, and he certainly had not solved the central problem of identifying the mechanism by which new species could arise. If Darwin had read more closely, he might have sensed that Wallace was getting close to an answer.

Wallace discovered a biological barrier running through the Spice Islands. There were two cleanly divided ecospheres; species on one side never seemed to mingle with those of the other. This was the same phenomenon that had puzzled Darwin in the Andes and the Galápagos: Why had God put different species in different places? What was the role of physical barriers? In 1858, during a bout of malaria and theorizing, Wallace recalled having read Thomas Malthus's essay on population. He thought about how diseases, accidents, war, and famine kept human overpopulation from exploding. The answer came in the following flash of insight:

> These causes or their equivalents are continually acting in the case of animals also; and as animals usually breed much more rapidly than does mankind, the destruction every year from these causes must be enormous. Why do some die and some live? And the answer was clearly, that on the whole the best fitted live. From the effects

of disease the most healthy escaped; from enemies, the strongest, the swiftest, or the most cunning; from famine, the best hungers or those with the best digestion; and so on.

Over the long term, changes of geography, climate, the food supply, and other factors would challenge organisms. The individuals best suited to the new circumstances would likely survive better, and after many generations this would produce new species. What seemed to be *adaptations* to their environments were simply the qualities that allowed them to survive.

In 1858 Wallace packed his new ideas into a paper called "On the Tendency of Varieties to Depart Indefinitely from the Original Type" and sent it off with a letter to Charles Darwin. The two had already been corresponding, and Wallace knew that Darwin was keenly interested in the laws underlying the development of species. In the letter he wrote that he hoped Darwin would find the idea original and helpful in solving the problem of species. He asked that his ideas be brought to the attention of Charles Lyell and other prominent scientists. Then Wallace packed his bags and boarded a boat bound for New Guinea, where he hoped to capture a living specimen of a rare creature called the Bird of Paradise.

Wallace's letter came as an enormous shock to Darwin. Although he had been working on evolution for almost fifteen years, he had not gone public with his ideas. Darwin had waited so long for several reasons. He had a good idea of what a furious controversy he would be setting off. The implications would be tremendous—for organized religion, but also personally. The Darwins' daughter, Annie, had died recently, and religion had been his wife's only consolation.

THE THEORY IS MADE PUBLIC

Darwin knew he could not offer ironclad proof for evolution. Like a lawyer lacking a "smoking gun," the best he could do was to collect as much circumstantial evidence as possible, show that

Who Really Discovered Evolution?

Things rarely turn out well when two or more scientists make the same discovery independently and almost simultaneously. Important discoveries not only help a scientist earn a place in history; they also have practical, personal benefits. They help researchers further their careers and obtain the freedom to pursue the questions they are most interested in. Charles Darwin and Alfred Russel Wallace might have waged a bitter fight over who should be named as the first discoverer of evolution. Instead, they developed a lifelong friendship and a close working relationship, evidence of each man's sense of integrity and honor.

Darwin helped get Wallace the credit he was due. Wallace did not try to grab the lion's share of the glory because he was a shy man; besides, he lacked the patience and discipline to spend years gathering thousands of facts to support the theory. At one point Wallace described Darwin as a "great General who can manoeuvre the largest army," whereas he considered himself good for "skirmishes or smaller campaigns"—unable to cope with the mass of details necessary to make a strong factual case for evolution.

Additionally, being known as the codiscoverer of a major theory was far more than Wallace had ever expected to accomplish. After years of virtually no recognition and a constant struggle to make a living—at one point, his entire livelihood depended on the survival of two rare birds—Wallace was suddenly propelled into the life of the scientific elite, as if he had suddenly become a movie star.

Both men recognized that their ideas owed a debt to many other scientists. After *On the Origin of Species* had been published, Darwin discovered that the idea of natural selection was not entirely new. In 1831, Patrick Matthew had described it in a book called *Naval Timber and Arboriculture;* a doctor named William Wells had discussed it in 1818, as had Henry Freke, an Irish physician, in 1851. Their books had gone almost unnoticed by the scientific community, and neither Darwin nor Wallace had read them.

Another person who had come close was Herbert Spencer, a journalist and philosopher who coined the term *evolution* and used it to talk about social progress nearly a decade before Darwin and Wallace presented their theories to the public. Spencer was familiar with the ideas of Lamarck, Erasmus Darwin, and Robert

Herbert Spencer *(Smithsonian Institution Libraries)*

Chambers and believed that the laws that ruled the natural world also applied to society. But he believed that changes in species and culture represented progress, and he had not proposed a theory of natural selection.

So many people were thinking along similar lines that if Darwin and Wallace had not come up with the theory, someone else surely would have. It might have been Thomas Henry Huxley (1825–95), one of Charles' relatives and a brilliant naturalist who had also traveled the world. When Huxley heard the theory he compared it to being struck by a lightning bolt—instantly recognizing it as the laws everyone had been looking for. He said, "How extremely stupid not to have thought of that!" He became one of evolution's staunchest supporters, calling himself "Darwin's bulldog," and helped win wide acceptance for the theory in the scientific community. He took on the battles that Darwin could not because of illness, his solitary nature, or the fact that there was too much scientific work to be done with the theory.

the theory was compatible with all the facts, and demonstrate that it could deal with problems and questions about life that no other theory addressed. He did this extremely thoroughly and methodically. But he had not rewritten his early outline because he was too busy collecting facts and he did not think anyone else was remotely close to his solution. Now Wallace was forcing his hand. If Darwin did not publish, two decades of work would be lost. If he did, he might look like a pirate who had stolen Wallace's work.

Depressed and unsure of what to do, Darwin asked Lyell for advice. His friend proposed a joint announcement of the theories at a meeting of a scientific group called the Linnaean Society. Darwin was unable to attend because another child, his infant son, had just died from scarlet fever. Wallace was still in Asia. Charles Lyell, with Joseph Hooker (back from the Himalayas), read the papers and added their statements of support. There was very little debate this first evening, because the ideas were too revolutionary and the audience needed time to think them through. But the world had now heard of evolution, and there was no going back.

Things would not remain quiet for long. The next year Darwin finished and published his book *On the Origin of Species,* which spelled out the theory in detail and presented an enormous amount of evidence in support of it. While the book itself did not take the theory to all of its logical conclusions—for example, it did not directly address the evolution of humans—the stakes were clear. Evolution surely applied to human beings, meaning it posed a threat to the Church's most treasured beliefs. Perhaps man was no longer a special creation of God, with divine characteristics. Perhaps he had descended from the apes, and was "just" an animal with some unique characteristics. Most of what scientists and the religious community had believed about life might be wrong.

3

The Theory and Society's Response

At the end of the 1830s *The Voyage of the Beagle* was a state-of-the-art travel book for the old way of looking at life, full of sharp observations about the features, lives, and interactions of species. But they remained single facts; Darwin was not yet able to perceive the grand system that unified them and made sense of them. *On the Origin of Species,* 20 years later, served as a guidebook to an entirely new world. Columbus's return from America was likely the last time that a discovery had such a deep and sweeping impact on people's worldview. Columbus had turned the world into a sphere and opened up new lands for European exploration; Darwin's work opened human nature and all of life to scientific exploration.

The theory touched off ferocious debates within science, at public gatherings, in the newspapers, and in churches—it was not easy to integrate into worldviews that had been held for thousands of years. On the one hand, evolution was so simple "that anyone could misunderstand it," as the philosopher David Hull has ironically put it, but in other ways it was quite complex. Science could not yet cope with some of the basic issues it raised. *On the Origin of Species* was a research proposal, a report on a vast work in progress, with the potential of becoming a complete theory of life. It offered the first scientific

explanation of the history and distribution of organisms and replaced the notion of intelligent design with a set of natural laws.

Darwin purposefully left out two issues that everyone was
interested in—the beginnings of life itself and the origin of humans (which he would address in another book, *The Descent
of Man,* published many years later). He avoided the first topic
because scientists did not know enough about microorganisms,
the simplest and most ancient forms of life, to classify them or
understand their evolution. As for the second, Darwin had not
yet studied humans and primates with the intensity he had examined barnacles and other animals. Waiting to address human
evolution might give people the time to get used to the theory,
and time for scientists to collect further evidence. Wallace, on
the other hand, did not hesitate to claim that human beings had
arisen from earlier forms of life and had been shaped by natural
selection. Ironically, he later changed his mind and began to
believe that certain aspects of human beings could not be explained by natural selection. Darwin believed that Wallace was
wrong about these questions, and nearly all of today's scientists
agree. To understand why, it is necessary to look more deeply
into what evolution said, what it did not, and what modern science has made of the theory.

A RECIPE FOR NEW SPECIES

Evolution is based on a few straightforward principles. Its complexity comes from how these principles work together, the in-
tricacy of life itself, and the fact that evolution nearly always
takes place in environments where thousands of species are
constantly interacting. The list below draws together the main
points Darwin mapped out in *On the Origin of Species.* The real
work of understanding evolution lies in appreciating the pro-
found meaning of each of these points, its implications, and
how the mechanisms work together.

The following ideas form the main theses of *On the Origin of
Species* and all extend from the concepts of variation, heredity,
and selection:

- Heredity. Organisms pass along most of their characteristics to their offspring.
- Variation. Within species there is a great deal of variation; individuals are neither identical to their parents nor to each other. Some of these variations are passed along to the next generation through heredity.
- Acquired characteristics are not inherited. An organism's makeup is the product of hereditary information from its parents and variations from other sources (today the reasons are known: Mutations and other types of errors creep in as *DNA* is copied). Learning or lifetime experiences may be passed along by teaching, but they do not directly enter the physical hereditary material that is passed along to an organism's offspring. The only exceptions that have been found are certain types of viruses or bacteria that inject new DNA into the genomes of their host. If the change affects egg or sperm cells, it may be passed down to the next generation.
- Reproductive advantages. Some individuals have more healthy offspring than other members of their species. A plant may produce more seeds, or an animal may have larger litters. Or a feature may help the organism survive long enough to reproduce. If an organism has more offspring, more of its *genes* will appear in the next generation. If its offspring also reproduce more, and this cycle of bias repeats long enough, the genes from one "family" will eventually dominate an entire species.
- Natural selection. In the long run, every successful species produces more offspring than can survive to adulthood. Environmental factors such as climate and predators play an important role in determining which will survive. Selection can be negative (for example, by eliminating organisms that have difficulties surviving and reproducing) or positive (increasing the number of descendants a particular organism has). Selection is a result of the interaction of the organism and the environment. For example, camouflage is often a good survival strategy, but it is only successful through a

combination of the organism's own color, background colors in the environment, and the sharpness of predators' eyes.

- Adaptations happen when natural selection works on variation that already exists within a species. Organisms do not change their hereditary material in order to survive; a characteristic already has to be present to undergo selection. Over time, if the environment remains stable (for example, if no new predators arrive), selection leads to a better "fit" between an organism and its environment.

- Natural selection works constantly, from the moment of conception to death. An organism has to survive each stage of its development as an embryo, birth, childhood, adolescence, and adulthood to pass along its genes to the next generation. After it has finished reproducing, it may help younger members of its species to survive, so natural selection may even influence organisms that are too old to reproduce. But in most cases its influence stops when organisms are no longer fertile. As a consequence, humans and other mammals have evolved immune systems that protect them from many threats that they face in the early parts of their lives—such as infectious diseases—but do not have defenses against "old-age" diseases like Alzheimer's.

- Natural selection can apply to any level within an organism, ranging from events within cells to the formation of tissues and behavior, if these features are hereditary. Even *altruistic* behavior (self-sacrifice) can undergo positive selection if it improves the chances that an organism's genes will survive in the next generation.

- Competition. Some varieties dominate an entire species very quickly, completely replacing other forms. This happens if they offer such major advantages to an organism's success and reproduction that over just a few generations, the genes that cause them spread through the entire population. This should not be confused with deliberate human competition. Organisms may not be aware of it or want it to happen. And varieties that have less of an advantage do not always become extinct; they may make minor contributions to a species's *gene pool* for a long time, or even develop into new species.

- Isolation plays an important role in the development of new species. If two groups from the same species become separated, they will undergo different random variations and thus experience different types of natural selection. If the two environments are quite different, the pressures of selection will affect them in very different ways. But even when environments are almost identical, such as in the Galápagos Islands, populations that become separated usually follow different evolutionary pathways.
- Natural selection is not the only influence on species change. Accidents, random events, and other factors also play an important role.
- Evolution is nondirectional. It does not necessarily lead to more complex or more intelligent creatures. It does not make organisms "better" in any sense except that in the long term, if the environment stays stable, they tend to have more offspring.
- Strictly speaking, species do not adapt "to" an environment; they adapt because of it. Adaptation does not occur within one generation, as a response to existing conditions. The current population of every species is the result of selection in the past. So the features of modern humans are largely the product of the environment in which the species originated, and the hunter-gatherer lifestyle that *Homo sapiens* practiced throughout most of its history.

Taken together, these principles are responsible for the origins of new species. They can also work on nonliving systems, such as chemical substances that arose on the early Earth. Anything that reproduces imperfectly in an environment with limited resources is subject to evolution.

Natural Selection versus Bottlenecks

Natural selection operates over many generations to give rise to new species. Over time the "fitness" of organisms increases because the environment favors those that are better equipped to survive and reproduce. Sometimes, however, an organism survives for unusual reasons, or simply because it is lucky. When an individual or a small group survives and the rest of the species dies out, due to a catastrophe or something else, scientists say that there has been a *bottleneck*. This is an unusual situation because it determines the entire future fate of the species. All of the characteristics of the survivors will be passed along to the next generation, not only features that would normally undergo positive or negative selection. So bottlenecks are usually exceptions to the way that natural selection normally works.

The following are some examples of how this happens:

1. About 75 years ago, yeast cells were taken from a moldy fig and brought into the laboratory to be used in the production of beer. Over decades, scientists repeatedly selected the cells that did the best job at fermentation and discarded the rest. Thus the yeast cells most commonly used in today's laboratory experiments have undergone many bottlenecks through "artificial selection."

2. In the late 18th century, a tsunami washed over a very low-lying Pacific island called Pingelap, killing about 900 of the island's 1,000 inhabitants. Most of the survivors belonged to the royal family. Centuries of inbreeding had given many of them a rare genetic form of color blindness called *monochromatism*. Their eyes have no cone cells, which

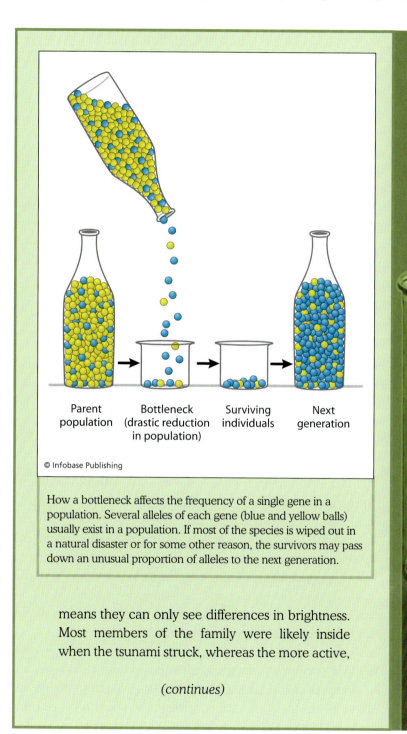

Parent population   Bottleneck (drastic reduction in population)   Surviving individuals   Next generation

© Infobase Publishing

How a bottleneck affects the frequency of a single gene in a population. Several alleles of each gene (blue and yellow balls) usually exist in a population. If most of the species is wiped out in a natural disaster or for some other reason, the survivors may pass down an unusual proportion of alleles to the next generation.

means they can only see differences in brightness. Most members of the family were likely inside when the tsunami struck, whereas the more active,

(continues)

(continued)

healthy population was outside and most were killed. Today, monochromatism is extremely rare in most of the world—occurring at a rate of about one in 30,000 people—but about 10 percent of the population of Pingelap suffers from the condition. Oliver Sacks recounts this fascinating story in his book *The Island of the Colorblind*.

3. Another bottleneck in human history occurred when the early ancestors of modern humans left Africa to settle the rest of the globe. Genetic studies of humans around the world show that they all descend from a single woman—nicknamed "Eve" (not to be confused with the character in the Bible). Thus we all carry the descendants of her genes.

MISINTERPRETATIONS OF EVOLUTION

Darwin and his contemporaries spent years trying to help scientists and the public understand the theory, but many people found it difficult to cope with a completely new way of thinking about life. They hoped that evolution was wrong, or that it could somehow be aligned with their old beliefs. Attempts to make the old and the new fit together caused confusion and often reflected wishful thinking, for example, by trying to bend the theory to give humans a special status in nature. Many of these misinterpretations have survived until today in some form, and clarifying them can help explain the theory.

The list below gives some of the most common misconceptions and shows which principles they violate. Darwin addressed many of these issues in his books, but some of his answers have been updated to account for the enormous amount that has been learned about evolution over the past 150 years.

1. "Humans are the high point or the goal of evolution."

Some people claim that "evolution happened in order to produce humans," or that "evolution had to produce intelligence," but evolution is nondirectional. It does not necessarily produce more intelligent creatures, or larger and more complex ones—today's animals are often smaller and even genetically simpler than their fossil ancestors. Sponges have existed in some form for nearly a billion years without developing a brain or getting much smarter. And one of the most intelligent species that ever lived, *Neanderthal* man, became extinct without leaving any intelligent descendants. Evolution did not "intend" to produce humans, any more than it intended to produce bacteria or insects, although in sheer numbers these are among the most successful creatures alive on Earth today.

Once certain types of intelligence existed, they could become subject to natural selection like every other feature of organisms. The ability to reason and its products—science and technology—may certainly help humans to avoid disasters that would destroy other species, but it is also possible that humans will one day become extinct without leaving intelligent descendants.

2. "If you were to put all the components of a cell in a test tube and shake it for the entire lifespan of the universe, you would never get a bacterium, so life could never have evolved from inorganic substances." Jonathan Wells, of a creationist organization called the Discovery Institute in Seattle, Washington, put it this way: "The problem of assembling the right parts in the right way at the right time and at the right place, while keeping out the wrong material, is simply insurmountable."

The flaw in this argument can best be demonstrated by an analogy. In 2002, Andrew Whittaker won $315 million in a "Powerball" lottery. Although the odds of a particular person holding the winning combination of numbers are often over hundreds of millions to one, lotteries are often won if enough people buy tickets. On the other hand, wagering that a particular person will win is a bad bet. This situation is similar to the hypothetical test tube described above. It is extremely unlikely that life will evolve in one particular way, to produce a bacterium; that process involved billions or trillions of steps, over a huge span of time, and has surely only happened once in the history of the universe. But most scientists believe that given enough time, and enough "experiments," natural selection is likely to create some form of life out of inorganic substances. It is impossible to predict what it might be like.

How many "experiments" happened on the early Earth? The "test tube" that produced life was the size of the entire planet. Each cubic inch (about 16.4 cm^3) held about 500,000,000,000, 000,000,000 molecules. There are about 254,358,061,000,000 cubic inches in a cubic mile (about 4.1 km^3), and today's oceans contain over 320,000,000 cubic miles (about 1,310,000,000 km^3) of water. Multiplying those numbers together gives the number of chemical experiments that were going on simultaneously, and there have been at least 4.5 billion years to try out combinations. (It did not take that long; evidence suggests that life arose within the first billion years of the Earth's existence.)

This does not mean that if it were possible to reproduce the exact conditions present on the early Earth, the results would be the same, any more than it is likely that the numbers of Andrew Whittaker's ticket would also win a second lottery. Too many random events were involved in the creation of life. Under any other set of conditions the first organisms would undoubtedly have been very different, and then evolution would have taken a completely different course. There do, however, appear to be some basic chemical recipes for life that can arise in many different environments. The next chapter describes experiments which have tried to replicate conditions on the early Earth, in hopes of shedding light on this process.

3. "The eye only functions because it has many complex parts that have to be perfectly attuned to each other. Natural selection could not have produced such a precision 'instrument' through a series of small steps, because none of the earlier forms would have worked."

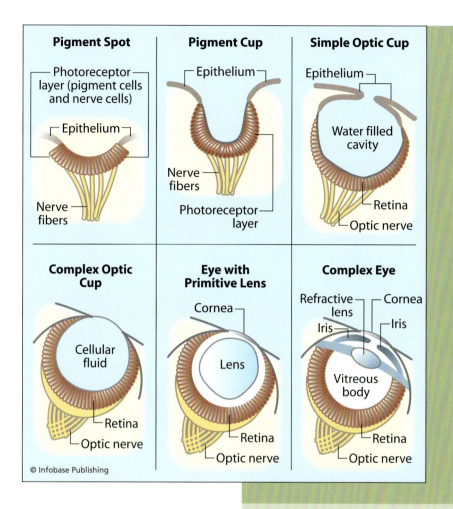

Examples of the many forms of eyes found in mollusks *(Source: Ridley, 2003)*

This claim is taken directly from the movement called natural theology, described in the first chapter, and it is incorrect for several reasons. First, it is simply not true. Some mammals, fish, insects, and other creatures alive today have eyes that are similar to those of our ancestors, and they work well enough to permit them to survive and reproduce. Even a very basic ability to perceive tiny differences in light and shade, which exists in many single-celled creatures, can give an organism an enormous advantage over others. Tiny freshwater worms called planaria have two simple cups of light-receptive cells in the head that act as primitive eyes. When they detect light, the worms swim away.

Very recent scientific work has demonstrated that eyes with very different architecture—from the simple eyespots of planaria, to the faceted organs of insects, to human eyes—are all based on a single type of "photoreceptor cell" that evolved once. They all use a similar *photopigment* molecule to transform light into electrophysiological signals. Photopigments are found in every animal and even bacteria, and over hundreds of millions of years, the cells that contained them gradually evolved into both eyes and nerves of the brain. New methods discussed in the last chapter of this book are allowing scientists to reconstruct some of the steps by which this happened.

Additionally, there is no reason why a single organism cannot undergo two or three mutually helpful mutations at the same time. The odds are against it, but those odds improve a great deal if the population of a species is large enough. The population of human beings, for example, is about 6 billion, and each person carries unique mutations, some of them in genes. Many simpler organisms reproduce millions of times faster, have smaller genomes, suffer more mutations, and have populations that are thousands or millions of times as large. So the natural world is an immense evolutionary "laboratory" that is capable of running an unimaginably large number of "experiments" at the same time.

Finally, even the proponents of intelligent design are well aware that an incredibly complex eye can arise from a single-celled organism, and that it can be built in a series of tiny steps out of much simpler structures. This happens during the development of every embryo, as a single cell divides and specializes to become an entire human being. This raises the issue of interesting parallels between the growth of embryos and evolution, discussed in chapter 6. Every argument about complexity ignores the fact that all incredibly complex embryos begin as single cells. If an eye can arise this way in a single lifetime, it can also arise over billions of years of evolution. The two processes are different, but one is no more unlikely than the other.

4. "Biochemical processes within cells are too complex to have evolved naturally, so they must have been designed by an intelligent creator."

Recent genetic evidence shows that even very differently formed types of eyes originated as a single cell in an ancient ancestor. *(top left)* Monkey eyes *(Kit Whitfield)*
(bottom left) Shark eye *(Wikipedia)*
(right) Eyes on dragonfly *(Photo by Geoff Woodard. Courtesy Eye Design Book)*

This is a paraphrase of the previous point, but it merits another look because it is an argument currently popular among supporters of the intelligent design religious movement. A recent version of the idea appears in a 1996 book by biochemist Michael Behe called *Darwin's Black Box,* which has received a

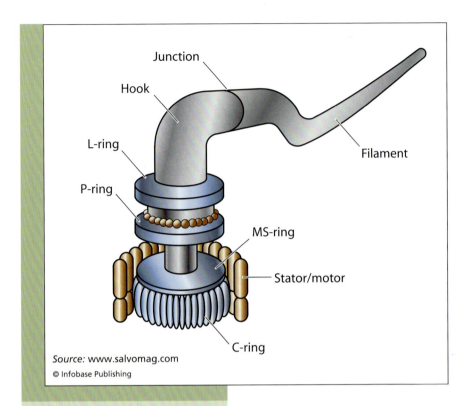

Junction

Hook

L-ring

P-ring

Filament

MS-ring

Stator/motor

C-ring

Source: www.salvomag.com

© Infobase Publishing

The hairlike flagellum that propels many species of bacteria has been called an example of "irreducible complexity" by proponents of intelligent design. However, related forms of the molecules that make up this structure are found in many other types of cells, where they perform a variety of functions.

lot of attention because Behe is a scientist and the language and style of his arguments themselves sound scientific. Behe claims that some processes in cells are "irreducibly complex"—in other words, they could not have evolved from a smaller, simpler set of components.

While Behe is an expert biochemist and his arguments sound compelling, many other scientists have carefully studied his opinions and determined that the claims are not supported by evidence. First, the processes he describes (such as blood clotting, the creation of antibodies, and the operation of the hairlike flagella that propel bacteria around) are carried out by Lego-like machines of proteins. Their elements are found in many species in a variety of forms. Those forms are often simpler than in the flagella, and they often serve different functions. This is precisely what evolution predicts. Nature does not invent new sets of parts as every new species arises; instead, it acts on what is already there. Sometimes new functions arise through new combinations of parts.

An analogy to Behe's argument might be this: Suppose that a tiny automobile made of Legos is found rolling down a street. After a careful study of the car a person finds that if any parts were removed, it would not roll. Behe's analysis is like saying this is the only possible way that Legos could be put together to make an automobile—or to make anything else, for that matter. Natural selection, on the other hand, supposes that there might be trillions of ways to put Legos together, some of which would roll downhill. Suppose a person standing downhill collected all of the structures that arrived at the bottom (playing the role of "natural selection"), then started improvising on their designs. By adding and taking away random pieces (also made of Legos) he would likely create other successful rolling designs. A lot of them might be "irreducibly complex"—in other words, taking away a single piece would stop them from rolling. (In natural evolution there is no one standing around making the car; it reproduces itself, and makes small mistakes all the time.)

Ken Miller, a professor of biology at Brown University in Rhode Island, provides a very thorough analysis of Behe's claims on his Web site, which is listed in the Further Readings section of this book.

5. "There has not been enough time since the formation of the Earth for evolution to have produced incredibly complex creatures like human beings."

This idea is probably an echo from the late 1860s, when physicists mistakenly estimated that Earth might only be about 100 million years old. They had made three false assumptions: that the Sun developed out of a gas cloud that condensed at an equal pace over time, that the Earth had once been a ball of liquid lava which cooled at a constant rate, and that the salt in oceans came uniquely from rivers, delivered at a steady rate. Their calculations worried Darwin, who believed that evolution happened at a slow pace. He need not have been concerned; they had made a mistake. The Earth is over 40 times that old.

Estimating how long evolution takes depends on knowing at least three things: the age of the Earth, the time at which life originated, and the rate of evolution in each species since life began. Scientists are closest to knowing the first: Geologists and astrophysicists believe that the Earth formed about 4.5 billion years ago, based on careful studies of radioactive characteristics of ancient rocks and meteorites. The second date will probably be impossible to determine very accurately, but many scientists believe that there was already some form of life on Earth nearly 4 billion years ago.

Figuring out the third item, the pace of evolution, is a fascinating challenge which may become a bit easier now that researchers can study the complete genome sequences of many different organisms. Comparing their genetic codes provides a sort of "molecular clock," and recent studies have combined this evidence with the fossil record to study the origins of humans, penguins, and other birds, moths, horses, and many more species. The clock runs at different rates for different species; organisms such as flies that mature and reproduce very quickly also evolve more rapidly. Everything that has been discovered in DNA sequences supports evolution's claims. This, too, is discussed in more detail in chapter 6.

6. "The fossil record does not confirm evolution because there are *missing links.*"

The idea that if humans descended from the apes, somewhere there ought to be a creature that is half man, half ape, is a misinterpretation of evolutionary theory, which states only that humans and other primates arose from a common ancestor. What kind of a creature was it? It most likely looked enough like a human and enough like a chimpanzee that both species would have welcomed it into the family tree. But to think of it as half man and half monkey is like thinking that the great-grandmother of Sharon and Lisa is half Sharon and half Lisa. Just as it might not be possible to look at the children and then pick their great-grandmother out of a lineup of suspects, it might not be possible to recognize a fossilized common ancestor of humans and apes.

Many fossil ancestors will never be found because only a tiny proportion of the animals that have existed on Earth became fossilized, and only a tiny fraction of those have been discovered so far. Another complication is that it is usually impossible to determine where and for how long common ancestors existed. In some cases this might have been a very short period of time, because each time a group was split by migrations, geography, or other factors over enough time, the separated groups quickly began to evolve in different ways.

An analogy might be to look for a common ancestor of two women ("Gabi" and "Karen") on the passenger list of the *Mayflower*. That was a small group, like the number of hominid fossils that have been found. Many current-day Americans descend from that original group of settlers. But many do not. It would be illogical to claim that since no common ancestor of Gabi and Karen was aboard the ship, they do not have one.

In spite of these issues, paleontologists do continue to unearth "missing links." In 1994 intelligent design advocate Michael Behe (mentioned in point 4 above) stated that if whales descended from a land-living mammal, there ought to be "transitional forms." He cited the fact that none had been found as a serious challenge to evolutionary theory. Within just months of his challenge, scientists found three separate fossils bearing features of exactly such transitional forms. The same thing has happened regarding the evolution of the ear. The small bones that permit humans to hear are remarkably similar to bones in the jaws of reptiles and earlier life forms. Critics of evolutionary theory have often denied that bones serving one of these functions could develop into the other. But in 2006 Zhe-Xi Luo and his colleagues at the Carnegie Museum of Natural History in Pittsburgh discovered that the bones of an ancient mammal called *Yanoconodon allini* represented a transitional structure between ancient jaws and modern ears.

7. "Natural selection would favor selfish individuals, and therefore human qualities like generosity and altruism could not have arisen through evolution."

Sometimes behaving very selfishly might give an animal the best chance to survive and pass along its genes. At other times this could even harm its chances, which has been demonstrated using a type of mathematics called *game theory,* discussed in chapter 4. Extremely selfish and aggressive behavior might not always pay off for an individual; it might well lead to reprisals from other members of the species. And a creature that never shares its food and does not care for its family may survive, but that does not mean it will have the most children that survive long enough to reproduce. In the case of humans, children are dependent for so long that they would not survive without generosity and altruistic acts on the part of their parents and other relatives. This social behavior on the part of humans and other animals may become the subject of natural selection. It is what author Richard Dawkins calls the extended *phenotype,* explained in more detail in chapter 4.

Successful and frequent reproduction—the cornerstone of evolution—is not all about competition for food; the real question is what ancestors had to be like in order to pass along their genes. Dawkins and Bill Hamilton put this question into an entirely new perspective by claiming that evolution operates on the basis of *selfish genes*—genes that increase the likelihood of their own survival in various ways. The best strategy for a selfish gene may be to help shape an animal that shares and sacrifices itself for its offspring.

8. "Animals cannot make art or music, and these activities would not have had any survival value, so they could not have arisen through evolution."

The brain did not evolve in order to compose Bach's fugues or Mozart's symphonies, or even to produce language, any more than the hand evolved in order to move a computer mouse. However, a wide range of creativity and problem-solving skills surely helped our ancestors survive and reproduce. Each animal's brain was a little different, just as every human being's brain has tiny, unique features. A primate's brain did not come with a user's manual telling what should and should not be done with it. But with slight changes in structure (accidents of heredity) came new types of behavior, and some of them were advantageous when it came to passing along genes. It is a fabulous accident that the brain that helped human ancestors survive long ago, in a much different environment, also turned out to be capable of doing mathematics, art, language, science, and the other skills that people value. These skills are examples of something that is true for nearly everything in nature: A structure selected for one thing becomes capable of completely new, additional functions through random variations.

If some artistic or cognitive skills are due to unique features in a person's brain, and if these features are hereditary, they may be subject to natural selection in the future. If, for some reason, rock musicians or math geniuses suddenly begin to have more children than the rest of the population, future generations will compose better music and be better at math. On the other hand, if being a "couch potato" is a hereditary trait, and couch potatoes reproduce more, human evolution may take a much different direction.

9. "Evolution has not been proved because it is only one theory among many, and even scientists disagree with each other about the theory."

The word "theory" means different things in different contexts. In everyday language it often refers to an opinion or a personal philosophy about something fairly trivial. In science it has a much stronger meaning—a theory is usually considered to be a collection of related, logical principles which explain a large number of facts. In daily life, people sometimes hold onto personal opinions although the facts seem to contradict what they believe. Science, however, has to abide by facts.

Scientists are creative and competitive people, and they enjoy overturning theories if the facts do not support them. (Anyone who does so successfully can look forward to fame, more money and freedom to do interesting research, and sometimes a Nobel Prize.) If evolution had not occurred, it should be easy to find contradictions to the theory—the discovery of a fossil of a modern human that was 100 million years old, or seeing that human DNA resembled that of a bacterium more than that of a fish. Such facts would force scientists to reevaluate evolution completely. In fact, some of Darwin's ideas have been discarded because of discoveries over the past 150 years, but not the core of the theory.

Although today's scientists almost universally accept evolution, there are still controversies over the details. Some of the main issues are discussed in chapter 5.

10. "Everything about humans is the product of natural selection, so it is natural that some people should be wealthy and others poor."

The idea that humans evolved through natural forces suggests that human behavior and social structures are also "natural" and are likely to be influenced by heredity and natural selection. Evolution suggested that some humans might be more "fit" than others, and in the 19th century many people assumed that power and wealth were forms of human fitness, the way that strength and speed helped animals escape from predators. The comparison was based on the social values of the time (the fittest ought to be the best, or the most admirable) rather than evolutionary principles (the fittest are those that reproduce the most successfully). Equating success in society with success in evolution, trying to draw parallels between social and biological evolution, or thinking that different races showed different degrees of fitness all go far beyond Darwin's theory. This theme is explored in the rest of this chapter.

11. "Since evolution cannot explain everything, it did not happen."

Every day researchers discover things that they cannot explain; they look forward to such moments. An inexplicable result is one of the best outcomes of an experiment, and when it happens it does not mean that a researcher has found evidence of miracles, God, or the account of creation given in the Bible. An analogy can make this clear: Airplanes appear to defy the laws of gravity. Obviously they do not; there is a scientific explanation for their behavior. But suppose a Stone Age proponent of intelligent design saw an airplane. Since he could not explain it, he would feel justified in calling it a miracle. He might even see it as proof of the existence of God. Not all advocates of intelligent design are "creationists," but many are, and this person might even present the airplane as evidence of the account of creation given in the book of Genesis. By claiming that any other opinion was misguided or the product of evil, he might stop his friends from wondering how a machine could fly. And then there would be no reason to try to build one. Doing so might take years, centuries, or even longer, but it would be a worthwhile project, and along the way many other interesting and useful things would be discovered.

Over the years proponents of "intelligent design" have claimed that gaps in the fossil record, complex biochemical processes, and other mysterious phenomena call evolutionary theory into question. These observations have usually had a contrary result: By challenging scientists to find answers, they have led to the discovery of new facts and mechanisms which support evolutionary theory.

INITIAL REACTIONS TO EVOLUTION

The publication of *On the Origin of Species* initially polarized the scientific community and society. Evolution had many strong supporters, such as Joseph Hooker, Charles Lyell, and Thomas Henry Huxley (who later called himself "Darwin's bulldog" in reference to his fierce support of evolution). It was also received favorably by thinkers in other fields, such as John Stuart Mill (1806–73), and Henry Fawcett (1833–84) at the University of

Cambridge, who were starting to think about what the theory might mean for economics and mathematics.

Strong opposition came from figures such as Richard Owen (1804–92), the first director of the Museum of Natural History in London. Like Georges Cuvier, Owen believed that animals' bodies were perfectly fit to their environments and were the product of intelligent design. Another scientist who was violently opposed to the theory was Adam Sedgwick, Darwin's old geology teacher; he followed the hard line of intelligent design, saying that the purpose of science should only be to bring people closer to God. The *Beagle*'s captain Robert FitzRoy, now a deskbound admiral in charge of setting up the world's first weather service, was so upset that he wrote Darwin an emotional letter and published an anonymous antievolution article in the newspaper.

Some religious leaders responded in the same way that they had accepted new findings in astronomy and geology. Evolution said nothing about the origins of the universe, and they could cope as long as it left room for God to have set things in motion, including the natural laws that would take care of the rest. But many clergymen were forcefully opposed, and they condemned Darwin in sermons, stating that the theory directly contradicted the Bible and called into question the existence of an immortal soul.

The first major confrontation came in a huge public meeting in Oxford in 1860 that was attended by about 700 people. Once

T. H. Huxley, "Darwin's bulldog." From *The Outline of Science*, edited by Arthur Thomson. *(G.P. Putnam's Sons, 1922)*

again Darwin could not attend; he had suffered a severe relapse of a stomach illness. Robert FitzRoy stood up, waved a Bible, and said he regretted ever haven taken Darwin on board his ship. After delivering a long speech, Samuel Wilberforce (1805–73), bishop of Oxford and the figurehead of the antievolutionists, turned to Huxley and asked ironically whether he descended from an ape on his grandfather's or grandmother's side.

Huxley could not resist making a sharp retort. Later he was quoted as having said, "If I would rather have a miserable ape for a grandfather or a man highly endowed by nature and possessed of great means and influence, and yet who employs those faculties for the mere purpose of introducing ridicule into a grave scientific discussion—I unhesitatingly affirm my preference for the ape." These probably were not his exact words, but whatever he said, the battle lines were now drawn for a major confrontation. Science and religion would be divided over the issue, and the debate would spill over into politics, education, and social philosophy. Although evolution was strictly a scientific theory, like the theory that the Earth was round, everyone felt the right to attack it on any grounds whatsoever. The topic had the feeling of a public brawl, open to all.

HERBERT SPENCER AND THE EVOLUTION OF SOCIETY

Evolution appeared at a time when rapid technological progress was dramatically changing society. Factories were speeding up the production of goods and making them less expensive; steamships and the railroad were revolutionizing travel, and the telegraph permitted almost instant communication over long distances. The gap between "advanced" and less developed societies was widening. People wanted to understand what these changes meant, but the only tools they had were philosophies of culture that were still heavily based on religious ideas. Governments used moral or religious arguments to rationalize their existence, structure, and behavior (often saying one thing and doing another).

Several years before Darwin published *On the Origin of Species,* a journalist named Herbert Spencer (1820–1903) proposed that the laws governing the biological world also ruled society and human culture. Like Darwin and Wallace, Spencer had been deeply influenced by Thomas Malthus's attempt to look at society in a scientific way. Spencer had already concluded that animals, human culture, and the entire universe were evolving. He was the first to use the word *evolution,* although he was strongly convinced that it was a process of improvement. He also coined the phrase *"survival of the fittest,"* by which he meant that the strong, the most advanced, or otherwise the best would prevail. When Darwin's book appeared, Spencer was excited because he believed it provided a firm biological foundation for his own ideas. The theory should make it possible to prove that humans and society were part of the natural world, and thus subject to natural laws.

Spencer was a progressive, very influential thinker. He strongly promoted the equality of the sexes at a time when women had no right to vote, were rarely permitted to get an education, and were allowed to practice very few professions. He was extremely critical of governments' use of morality to justify their domination and colonialization of distant lands. "Sacrifices and prayers have preceded every military expedition. . . . 'God is on our side' is the universal cry . . . the Spaniards subdued the Indians under plea of converting them to Christianity . . . and we English justify our colonial aggressions by saying that the Creator intends the Anglo-Saxon race to people the world."

His ideas were attractive to Wallace, who had always been concerned about social injustice and hoped that the theory he had helped discover might change people's thinking about society in positive ways. Wallace had begun to think that natural selection might somehow be biased toward improving species. If so, he thought, it might shorten the time that had been necessary to create complex animals from simpler ones and produce human beings. Wallace was trying to find a way to show that evolution could take shortcuts because of a hypothesis suggesting that the Earth might be much younger than geologists had believed. He encouraged Darwin to use Spencer's phrase "survival of the fittest." Darwin was uncomfortable about this idea, but he did not want the scientific community to become divided by details. As a result he made some compromises about several issues, including this one, in later editions of *On the Origin of Species.*

Spencer used "survival of the fittest" very broadly when talking about society. Nature eliminates imperfect beings, he believed, including the old, the weak, and the handicapped. In an article called "Progress: Its Law and Cause," which appeared in 1857 (two years before the publication of *On the Origin of Species*), he wrote, "Whether it be in the development of the Earth, in the development in Life upon its surface, in the development of Society, of Government, of Manufactures, of Commerce, of Language, Literature, Science, Art, this same evolution of the simple into the complex, through a process of continuous differentiation, holds throughout." Left on its own, without governments to intervene, human society would progress by favoring stronger and healthier individuals—not necessarily the rich, because he realized that poorer social classes were not really responsible for the conditions in which they lived. On the other hand, some of these people were clearly "unfit" through idleness or incompetence, and when they starved or suffered, Spencer believed this was a natural process.

Many people who read Spencer's books saw the parallels to Darwin's account of biological evolution without understanding the important differences. Spencer's "evolution" referred to his desire to see a moral and technological improvement of human society, but its function was not to create new species. Darwin did not think this way at all, and he was uncomfortable with most attempts to apply evolutionary ideas to culture. He had a strong sense of the limits of the theory and of science in general. Many people were not able to draw a clear line between the two men's ideas, and this would have an important impact on the relationship between biology and society over the next century.

It is not hard to see how Herbert Spencer's ideas could be turned against the poor, the sick, or groups that were considered somehow "unfit" by those in power. Spencer was against charities and donations to the poor on the grounds that they ran against the principles of selection and promoted the survival of the "unworthy." "The quality of a society is physically lowered by the artificial preservation of its feeblest member," he wrote. Helping "defective" people survive could harm society and perhaps even the human race. He hoped that Great Britain would enact laws to restrict "degenerate classes." That did not happen immediately, because Spencer's ideas on issues such as war, colonialism, and the church had made him unpopular in England. He was much more popular in the United States. Soon some of his ideas would be put to use there and elsewhere on a large social scale, and when Spencer's intentions, too, were misinterpreted, the stage was set for things to take a terrible turn. Members of the Darwin family played a role in this.

FRANCIS GALTON AND POSITIVE EUGENICS

Darwin's cousin, Francis Galton (1822–1911), was a brilliant man who inherited a fortune when his father died. He decided to use it to become a "gentleman scientist." He traveled to remote areas of Africa and sent back scientific reports; back at home, he measured human characteristics and aspects of society and studied them using statistics. One of his accomplishments was to convince the police that fingerprinting could be turned into an exact science to identify criminals. He also carried out a scientific study of prayer, with negative results: He concluded that the royal family (who prayed frequently in public and were the subject of many other people's prayers) had no better health than others.

In one set of experiments, Galton set out to prove Darwin's hypothesis about the way heredity works. (He ended up doing exactly the opposite.) Darwin believed in an ancient Greek idea called *pangenesis,* which proposed that hereditary material was somehow transmitted in the fluids that males and females exchanged during sex. However, the role of cells in this process was not yet understood. Darwin proposed that the fluids contained reproductive "particles" called *gemmules,* created in the parents' bodies. They were normally dispersed throughout the body but collected in the sex organs. These somehow mingled during sex to create new organisms that blended characteristics from both parents. Other scientists recognized that such blending would "dilute" the effects of inheritance and likely make it impossible for favorable traits to gain a foothold in a species.

Francis Galton (*Smithsonian Institution Libraries*)

Galton reasoned that gemmules had to flow through the blood and tested the idea by transfusing blood from rabbits with fur of different colors into silver-gray rabbits. He expected that this would blend gemmules from the other rabbits into the hereditary material of the silver-grays, leading to offspring of mixed colors. When this did not happen, most scientists became convinced that Darwin's hypothesis was wrong. Darwin continued to hold to the idea, claiming that gemmules did not necessarily have to be transmitted in the blood. (He never learned of the work of Gregor Mendel, which would turn the study of heredity into the science of genetics.)

In 1865 Galton wrote "Hereditary Talent and Character," an article in which he claimed that intelligence and other behavioral characteristics were inherited. After studying numerous historical cases including that of his own family, from the eccentric genius Erasmus Darwin down to his cousin Charles, he concluded that about one-sixth of the offspring of famous people are likely to become famous themselves. This suggested that marriages between brilliant people would produce more such people and improve society.

Over the next 40 years, Galton promoted what came to be known as *positive eugenics,* an attempt to encourage the healthiest, most intelligent, and most talented people to marry. In Germany, for example, it was already a tradition for professors to marry the daughters of their colleagues. Such selective breeding would lead to improvements in the human race, Galton believed, although it might take a long time. But for many years he hesitated to encourage the government to draw any political conclusions from his work, because he wanted to be sure of his facts. He donated money to the University of London to establish a laboratory for the study of this subject. In England, positive eugenics became a mode, and Darwin's sons, notable scientists, politicians, and many others jumped on the bandwagon.

NEGATIVE EUGENICS

In the United States and a few other countries, eugenic thinking was taking a much more negative turn. In 1874, a physician and political reformer from New York named Elisha Harris (1824–84) became secretary of the New York Prison Association. He was a talented statistician and began to investigate a pattern he had noticed in the family names of criminals in country prisons. He traced an incredible number of "convicts, paupers, criminals, beggars, and vagrants" back to a family that had lived in Ulster County, New York, in the late 1700s. At the time of his study, six generations later, they had 623 descendants, many of whom became criminals. "In a single generation there were 17 children," Harris wrote. "Of these only three died before maturity. Of the 14 surviving, nine served an aggregate term of 50 years in the state's prisons for high crimes and the other five were frequently in jails and almshouses."

The famous judge Oliver Wendell Holmes (1809–94) and many others interpreted this to mean that criminal behavior was strongly hereditary. Others were more careful. A talented young statistician named Richard Dugdale (1841–83) followed up on Harris's study of the family (whom he gave the pseudonym "Jukes") with a very thorough investigation of his own.

After examining records and interviewing family members, employers, police officers, and many others, he concluded that the environment was more to blame in the family's tragedy than biology. What was being inherited, he wrote, was a pattern of neglect, abuse, poverty, other social factors, and physiological issues: alcoholism during pregnancies and sexually-transmitted diseases that affected unborn children.

Prisons could not reverse the Jukeses' fate, Dugdale wrote; if there was to be a solution, it would have to come through extensive social reforms that improved family members' lives. This would probably take two or three generations and would involve providing a secure environment, early education for children, and foster homes for orphans and children born out of wedlock. In later studies, Dugdale admitted that there was such a thing as a biologically inherited tendency toward crime, but it was very rare and could likely be corrected if environmental conditions were improved.

His conclusions went unheard. Dugdale died at an early age, and soon after his death his work was being misused to promote the idea that criminals breed criminals. The stage was set for negative eugenics; a public campaign to improve society by ridding it of "unfit" members. Some of America's leading progressive thinkers supported the movement.

Several social and historical factors contributed to the rise of negative eugenics. Darwin had avoided talking about human culture in terms of evolution, but his followers (and Wallace) did not hesitate to try to apply the theory's principles to society. This led to a catastrophe, partly because it was impossible to determine whether genes or the environment were responsible for undesirable traits. Supporters of eugenics believed that heredity played the dominant role.

For early evolutionary scientists it was clear that an individual organism is a product of both its heritage and its environment, but sometimes this is not clear from their writings. Random variation and heredity receive the most emphasis, partly because early Darwinists continually had to fight off Lamarckian misinterpretations that gave the power of species change to individual behavior and the environment. And many scientists sincerely hoped that their discoveries would improve the quality of life, which made it hard to separate their work from their moral values.

David Starr Jordan (1851–1931), the first president of Stanford University, became one of the most outspoken figures in America's negative eugenics movement. Unlike Dugdale, but like many scientists of the late 19th century, Jordan was convinced that heredity was far more important than the environment in shaping human behavior. Trained as a biologist, he began to believe that certain family lineages were "degenerate" and were spreading crime and poverty like a hereditary disease. Evolutionary principles suggested that the only way to truly reshape human beings would be to control their reproduction. Galton's positive selection was one way, but Jordan was more concerned about ridding humanity of its degenerates. The idea took an even more negative turn when he began to associate fitness with race and nationality. In 1907 he drew these themes together in a book called *The Human Harvest: A Study of the Decay of Races through the Survival of the Unfit.*

One of his concerns was the evolutionary impact of wars. Jordan believed that they had a huge, terrible influence on human evolution by killing off many of the strongest, bravest, and fittest young men in a generation. He actively campaigned for pacifism on speaking tours to Great Britain and Germany during the buildup to World War I. One of his stops was at the London Society of Eugenics, headed by Darwin's son Leonard, a major in the British army. Jordan found himself in front of a hostile audience; Major Darwin, for example, held that on the battlefield, the fittest would survive—a perfect example of "natural selection."

Back at home, Jordan obtained a chairmanship within the American Breeders Association and helped change its constitution to include a platform of eugenics. The association sponsored research into whether insanity and other mental diseases could be inherited. The committee established its Eugenics Records Office in the town of Cold Spring Harbor on Long Island, New York, which worked closely with a nearby biology laboratory headed by Charles Benedict Davenport (1866–1944).

Eugenics propaganda from "The New Virginia Law to Preserve Racial Integrity," by W. A. Plecker, Virginia Health Bulletin (vol. 16, no. 2), 1924 (American Philosophical Society)

These groups promoted public education about eugenics and began to promote the compulsory sterilization of "unfit" people.

As shocking as this sounds today, it was nothing new. In the 1890s, some physicians had begun to remove the ovaries of women with a history of "psychological problems," believing that this could improve their conditions. Castration or vasectomies were performed on males as a punishment for crimes or cures for "mental problems." These practices, the eugenicists said, were more humane than the death penalty, and they would have the added value of protecting society by ridding it of criminals. Many doctors were appalled, but the practices went on.

Dr. Henry Clay Sharp (1869–1940), a prison physician in Indiana, began promoting sterilization as a solution to insanity and "hereditary crime" around 1900. He petitioned the governor and the state legislature to pass a mandatory sterilization law. Despite protests from groups of physicians who claimed the practices violated patients' rights, the Indiana legislature passed the first compulsory sterilization bill and the state's governor signed it into law in 1907. A few other states had already seen similar bills but they had either been voted down or vetoed. With the Indiana law on the books, many others followed; by 1930, similar laws had been passed in 30 states.

Because there was no precise medical definition of "feeble-mindedness" or other illnesses mentioned in these laws, sterilization was frequently carried out based on very subjective criteria. Having a child out of wedlock, coming from a poor family, or engaging in "antisocial" behavior was sometimes enough to bring down a judgment of mental "unfitness" and sterilization. The courts often recognized the injustice of these practices and overturned the laws, claiming that they violated people's rights. Yet by the time Sharp died in 1940, over 35,000 people had been sterilized involuntarily in the United States.

The legal system had started down a slippery slope which would have horrible consequences. In the 1920s, Harry Hamilton Laughlin (1880–1942), superintendent of the Eugenics Records Office, turned his attention to immigrants. The United States was welcoming large numbers of immigrants every year, often poor people from Europe. Laughlin began a campaign to ensure that only "high-quality" immigrants would be accepted. This was pure racism in disguise, and Laughlin manipulated his studies with deliberate errors to support his conclusions. People opposed to immigration used his work as an excuse to justify their racist views.

The trends started to turn around in the 1930s. Scientists like Hermann Joseph Muller (1890–1967) showed that many of the so-called studies of human heredity of Sharp and Laughlin, which claimed that "feeble-minded" foreigners were arriving in the United States and producing feeble-minded offspring, were incorrect and ignored environmental factors like the inequality of women and huge differences in education and health among different social classes. Muller wrote, "There is no scientific basis for the conclusion that socially lower classes, or technically less advanced races, really have a genetically inferior intellectual equipment, since the differences . . . are to be accounted for fully by the known effects of the environment." Unfortunately, these voices of reason were not heard everywhere, and as the movement declined in the United States, it was on the rise elsewhere.

EUGENICS AND THE HOLOCAUST

The Eugenics movement found fertile ground in Germany, where many people were becoming obsessed with the idea of "race hygiene." Ironically, this stemmed from a movement in the latter part of the 19th century to improve public health. Rudolf Virchow (1821–1902), who discovered that cancer develops from malfunctioning cells, used his influence as one of Germany's most famous scientists to try to improve the health conditions of the lower classes. While promoting wider access to clean water and better sewage disposal methods, he began studying the heredity of hair color and using his observations to try to define the "Teutonic" or "Aryan" race. After a careful study he concluded that the idea of such a race was "Nordic mysticism" with no basis in science.

That did not stop his work from being followed up and misused by the German Society for Racial Hygiene, founded by Alfred Ploetz (1860–1939). Ploetz was not an anti-Semite, but as the society grew, it became full of members who deliberately targeted Jews as scapegoats for the country's problems after World War I, capitalizing on a tradition of anti-Semitism in Germany that stretched back hundreds of years. The society claimed that helping the poor and giving medical care to the "unfit" would dilute the Teutonic race and weaken the country. With the rise of Adolf Hitler and the Nazis, the road was paved to combine eugenics, nationalism, and the terrible postwar economic situation of Germany into a national program of sterilization.

Upon becoming chancellor in 1932, Hitler immediately proposed a mandatory sterilization law aimed at blind, deaf, mentally ill, and "malformed" people. He intended to rid society of its unfit members, a term that would soon expand to include

Buchenwald. "These are slave labor-
ers in the Buchenwald concentration
camp near Jena; many had died
from malnutrition when U.S. troops
of the 80th Division enter the camp."
Germany, April 16, 1945. Pvt. H.
Miller. *(National Archives and Records
Administration)*

non-Aryans. The Nuremberg
Laws of 1935 made it illegal
for Jews and non-Jews to
marry or have intercourse,
with the goal of making Germany more racially "pure."
In the mid-1930s the Nazis
began the systematic murder
of young children who were
judged to be retarded or seriously handicapped. With the out-
break of World War II, the targets expanded to include Jews,
homosexuals, people with mental problems, and anyone else
that the government regarded as "undesirable," including political dissenters. What began as a sterilization program to improve
the "German race" turned into the mass murder of millions.

Hermann Joseph Muller had tried to show that even when
promoted by people with good intentions, eugenics programs
were the puppets of economics and politics. There was no sci-

entific definition of race, and the science of heredity was in its infancy. Muller wrote, "An individual's total genetic worth is a resultant of manifold characteristics, weighted according to their relative importance, positively or negatively, for society. It is a continuous function of all of these combined, so that there is no hard and fast line between the fit and the unfit, based on one or a few particular genes." But in Germany, rather than regarding the evidence, politicians only cared to hear scientific results that supported their ideological positions. It is a lesson that should be carefully considered today, as ideological stances on evolution and themes like global warming are written into the platforms of political parties.

Today many people fear that genetic testing of fetuses prior to birth, combined with the legal right to obtain certain types of abortions, could become a modern form of eugenics. But modern science has shown that it is impossible to predict the quality of a person's life based on genes alone, and it has completely discredited the quest for "racial purity." The concept of race itself is vague and probably impossible to define in genetic terms. Humans said to belong to the same race are incredibly diverse; within a single race there is more genetic diversity than the average difference between people of different races. In a 2006 article called "Two Questions about Race," biological anthropologist Alan Goodman, who is president of the American Anthropological Association, shows how the focus of science in the genome age is human variation and diversity rather than race. "Americans and much of the world's population have been conditioned to think of race as a fuzzy jumble of behavior, culture, and biology: a deep and primordial mix of a bit of culture and a lot of nature," he writes. Race fails to explain 94 percent of human variation. "The degree of genetic variation between any two human groups is almost entirely explained by the geographic distance between them." This means that normally, a person is genetically more similar to his close neighbors (even those he would consider to belong to a different race) than to people of the "same" race, living far away.

The next chapter discusses how scientists have studied the spread of evolutionarily successful traits through populations.

This research shows that eugenics movements could not have achieved what they set out to accomplish, even if they had been motivated by the best intentions.

EVOLUTION IN SCHOOLS AND THE SCOPES "MONKEY TRIAL"

In 1900 few American pupils obtained more than a primary school education—which meant that families were far more responsible for shaping young people's ideas and opinions than schools. By the 1930s many more children stayed in school longer. By that time most scientists were strongly convinced that evolution had taken place and were including it in textbooks. This created a conflict between the research community, which expected evolution to be taught in schools like other accepted scientific principles, and fundamentalist religious groups who saw the theory as a threat to the values they wanted their children to learn. Things reached a high point in the mid-1920s when legislators in several states tried to pass laws forbidding the teaching of evolution in schools. Their first success came in Tennessee in 1925, when State Representative John Washington Butler wrote a bill making it a crime to teach evolution or "any theory that denies the story of the Divine Creation of man as taught in the Bible."

The American Civil Liberties Union (ACLU) was alarmed by the law because it required science teachers to promote a particular religious view in classrooms, a violation of the Constitution's separation of church and state and a threat to the beliefs of the many American children who were not Christians. The ACLU wanted to challenge the law in court, but this could only be done based on a specific case in which someone had broken the new law. The organization placed an advertisement in a Tennessee newspaper, offering to pay all the legal expenses of such a teacher. They found a volunteer in John Scopes (1901–70), a young man who had been a substitute for a biology class in Dayton, Tennessee. He was convinced by the scientific case for evolution and felt that pupils had the right to learn about it, particularly because the biology book used in the school had a chapter on Darwin and natural selection! How could it be against the law to teach what was in the school's own textbook?

A carnival atmosphere descended on Dayton as the town prepared for the trial, milking it for publicity. Journalists and celebrities began to arrive. The prosecution found a spokesman in the famous orator, fundamentalist Christian, and former Presidential candidate William Jennings Bryan (1860–1925). America's most famous trial lawyer, Clarence Darrow (1857–1938), took on the defense without charging a fee. Both felt that a great deal was at stake. The ACLU regarded the trial as a major test of individual rights versus the religious beliefs of the majority. On the one hand, democratic principles seemed to imply that the majority should be allowed to decide what was taught in schools. On the other hand, new scientific discoveries arose all the time, and it was felt that teachers should teach theories accepted by scientists—even those that contradicted religious beliefs. How else could religious freedom be ensured?

Darrow hoped to use the trial as a public education campaign for evolution by putting experts from biology, evolution, and even religion on the stand, but the prosecution undercut this strategy: The legal issue was not whether evolution was correct or not, but only whether Scopes had violated the law. When the judge agreed, Darrow's only option was to call Bryan to the stand as an expert on the Bible. Despite conceding that the Sun did not revolve around the Earth, Bryan claimed that the Bible should be regarded as the sole authority on matters like creation and did not need to be interpreted.

Scopes was convicted of having broken the law, and was ordered to pay a fine. The conviction was later reversed on a technicality, leaving no opportunity to pursue the issue in the courts. Soon several other states in the South passed antievolution laws.

The legal issue was not addressed again until 1968, when the U.S. Supreme Court took on the case *Epperson* v. *Arkansas.* Their ruling stated that all antievolution laws were unconstitutional because they violated teachers' rights and represented an attempt to promote religion in public schools. Since then, fundamentalist religious groups have made many attempts to subvert ruling after ruling of the courts. Common tactics are to try to redefine religious doctrines as some type of science (in the case of "intelligent design"), to portray evolution as some sort of subjective religious belief system, or to demand "equal time" for religious views, as if scientific theories were political campaigns.

In the meantime the vast majority of religious thinkers throughout the world have come to terms with evolution. As an example, the Catholic Church has stated that evolution and the Bible need not be incompatible—no more than the fact that the Earth goes around the Sun should challenge people's faith. In a 2006 book called *Creation and Evolution,* Pope Benedict XVI wrote that rejecting evolution in favor of faith and rejecting God in favor of science were equally absurd. The book promotes a theology of *theistic evolutionism*—in other words, evolution was the process by which God created life—and calls for people to stop making evolution a polarizing issue.

4

A Synthesis between Evolution and Genetics

As the 20th century began scientists still knew almost nothing about two of the cornerstones of evolution: heredity and variation. By the close of the same century, they had nearly finished reading the complete genetic code of humans, which provided deep insights into both aspects of the theory. Not only had researchers discovered the nature of the genetic material in our bodies—DNA in the nucleus of every cell—but also they now understood a great deal about the role it played in variation and evolution. These rapid advances started with the work of Gregor Mendel (1822–84) and then became the basis of genetics and molecular biology, which used techniques from chemistry and physics to read the history of evolution from genes.

It was not obvious to the first geneticists that their work was compatible with evolutionary theory. The problem occupied scientists for much of the early 20th century. This chapter traces the most important milestones along the way.

MENDEL'S WORK AND ITS REDISCOVERY

Darwin spent the 1850s and 1860s working out the details of evolution and performing experiments on plants and animals to try to get a handle on variation and heredity, which had him stumped. At the same time a monk named Gregor Johann Mendel, in a village called Brünn, on the European mainland, was making a huge leap toward understanding how heredity works. Over the course of eight years, Mendel carried out a series of extremely careful experiments in plant breeding to study how parent organisms pass along their traits to their offspring. He used peas because it was possible to control their pollination in a very precise way, eliminating contamination from other plants.

Mendel's methods drew on 150 years of studies of plant reproduction. The German botanist Rudolph Camerarius (1665–1721) had discovered that plants reproduced sexually; usually, pollen from the pistils of one plant had to reach the female sex organs of another to produce fertile seeds. He proved this by showing that if he cut off the pollen-bearing structures of male plants, the females would not make seeds. Twenty years later, Carolus Linnaeus repeated the experiment and showed that such sterile plants would once again produce seeds if he did the pollination himself. He also tried crossbreeding experiments, applying the pollen from one type of plant to another, sometimes obtaining hybrid species.

The method provided a way to control which plants were mating with each other, and Mendel understood that it could be used to study how parents contributed to the characteristics of their offspring. He chose peas because the *stigma,* the female sex organ, lies deep inside the plant and cannot be reached by the pollen of other plants. (In nature, a pea plant only pollinates itself.) He could remove the male pollen-bearing structure called the *anther* and then deliberately fertilize one plant with pollen from another. This gave him control over which parents contributed pollen to the seeds, and by planting them and watching what happened, he would see the influence from each parent.

Mendel succeeded where Darwin and others failed because he reduced the problem to its simplest form. He decided to investigate characteristics of the plant one at a time, rather than trying to understand how traits mixed together. He chose obvious characteristics like the color and forms of peas and sizes and shapes of other parts of the plant. In 1856 he began a series of precise experiments to measure their transmission from generation to generation. He kept records of the results and used his talents in mathematics and statistics to analyze them.

The experiments revealed several new facts. First, traits are composed of two parts, one from each parent. Secondly, males and females contribute equally to the characteristics of the offspring; a trait that comes from the father is not favored over one from the mother, or vice versa. Mendel discovered this by mating two plants with opposing characteristics, such as smoothness or wrinkledness, planting the hybrid seeds, and watching the offspring over several generations. All the peas in the first generation were smooth. But when he mated plants from this generation with each other, their offspring might go either way. On average, they bore three smooth pea plants for each wrinkled one.

This 3:1 ratio occurred over and over again, and Mendel suddenly understood what it meant—a characteristic like smoothness was built of two elements, or *alleles,* one from each parent. If the mother had two *dominant* elements (AA) and the father two *recessive* elements (aa), all of their offspring would have the type Aa and would show the dominant trait (roundness). If Aa plants mated with each other, their offspring could have the combinations AA, Aa, aA, or aa. The first three combinations would produce a smooth pea; only plants that had inherited two recessive elements (aa) would be wrinkled. The same proportions appeared with other characteristics such as green versus white color. The trials did not produce perfect 3:1 ratios (for the same reason that flipping a coin an even number of times often does not produce an even number of heads and tails). But over eight years Mendel carried out so many experiments that the final totals came very close.

Another of Mendel's great insights was that different traits did not interfere with each other: The smooth/wrinkled trait is

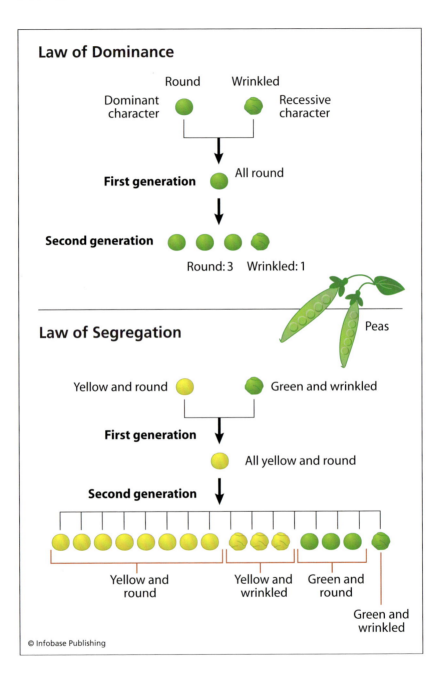

Law of Dominance

Round   Wrinkled

Dominant character   Recessive character

First generation   All round

Second generation   Round: 3   Wrinkled: 1

Peas

Law of Segregation

Yellow and round   Green and wrinkled

First generation   All yellow and round

Second generation

Yellow and round   Yellow and wrinkled   Green and round   Green and wrinkled

© Infobase Publishing

inherited separately from other features, such as whether peas are green or white. This discovery was only obvious because Mendel started by thinking that single characteristics might be *(opposite page)* Studies of patterns of inheritance for traits like color and shape in peas revealed the principles of heredity to Gregor Mendel. Above: The "law of dominance" reveals patterns of heredity for a single gene with one dominant and one recessive allele. Below: The "law of segregation" shows that the pattern of one trait does not influence the inheritance of another.

inherited separately—rather than assuming, the way Darwin did, that they were all merged in a liquidlike way. This influenced the way he designed his experiments and allowed him to prove that the traits moved from generation to generation in an independent way.

To test whether this hereditary behavior was unique to peas, he repeated the experiments on beans and other plants in the monastery garden. When he obtained the same results, he was confident enough to publish the work in the 1865 edition of the *Proceedings of the Natural History Society at Brünn* and present his findings in a public lecture.

Then Mendel had a bit of extremely bad luck. Knowing his work needed to be verified by another researcher, he sent off packets of seeds to a Swiss scientist named Karl Nägeli (1817–91), who had also been working on heredity. Nägeli had come to different conclusions because he had been working on a plant called hawkweed, which did not behave like peas. (In some cases, the plant reproduces *parthenogenically*—a type of cloning in which only genetic material from the mother is used to create offspring.) Mendel tried to reproduce the results from peas with hawkweed seeds and could not. He became depressed and uncertain of his results. While he continued to do experiments, he accepted an appointment to become abbot at the monastery and spent most of the last twenty years of his life immersed in the business of the monastery. He died before the importance of his work was recognized.

That would take several decades, but in the meantime another approach was starting to yield some basic information about heredity. Technical improvements in microscopes early in the 19th century gave scientists a much sharper view of the microscopic world, clear enough to show Matthias Schleiden (1804–81) and Theodor Schwann (1810–82) that plants and animals were composed of cells. In doing so they ushered in the modern science of cell biology. This led to the discovery of the role of sperm and egg cells in reproduction. Microscopists discovered threadlike structures in the cell nucleus that came in pairs (*chromosomes*). An egg and sperm carried the same number of chromosomes, which were mixed together when an egg was fertilized. Secondly, when cells divided, chromosomes were copied and the new "daughters" received a complete set. In the 1880s, Wilhelm Roux (1850–1924) proposed that these structures carried the hereditary material. August Weismann (1834–1914) followed up by predicting that egg and sperm cells had to be created by a special type of cell division that only gave them half-sets of chromosomes. That, too, turned out to be the case.

In 1900 Mendel's work was rediscovered simultaneously by three men in three different European countries. Working with poppies, the Dutchman Hugo de Vries (1848–1935) rediscovered the 3:1 ratio and proved that the principles held for 20 other plants, drawing conclusions almost identical to those of Mendel. In Germany, Carl Correns (1864–1933) found the same type of behavior in corn. An Austrian, Erich Tschermak von Seysenegg (1871–1962), had begun to work on peas after reading about some of Darwin's experiments and did a study that almost exactly reproduced the work of Mendel. All three studies were published in 1900. Later, each author claimed to have learned of Mendel's experiments only after reaching the same conclusions independently.

William Bateson (1861–1926) encountered Mendel's name when he read de Vries's paper, "Concerning the Law of Segregation of Hybrids," while on a train to London. He quickly rewrote the lecture he planned to give there, understanding that Mendel's principles might provide a way of linking heredity to evolution. Later he used chickens to test whether the laws also worked in animals and found that they did. More work turned up traits that did not obey the 3:1 ratio; particular colors in flower petals, for example, sometimes appeared only once in 16 flowers. Bateson immediately understood the significance of this: Color depended on the combination of two traits. A flower had to inherit two recessive genes to become a specific color, and the chance of that happening was 1:16.

Bateson had Mendel's paper translated into English so that it could be read by many more scientists, and in 1902 he published a book called *Mendel's Principles of Heredity: A Defence.* It caused an immediate sensation. The same year Archibald Garrod (1857–1936) showed that a human disease, a form of arthritis called *alkaptonuria,* followed inheritance patterns for recessive traits. This meant that human heredity also obeyed Mendel's laws.

Another of Bateson's contributions was to invent names for some of the new concepts in heredity. He called the entire field "genetics." (The word *gene* was contributed by a Danish botanist named Wilhelm Johannsen.) In most organisms, each gene appeared in two copies (one from each parent), which Bateson called alleles. If the alleles were identical—for example, if a pea carried two copies of the recessive gene for wrinkledness—he called the organism *homozygous* for that trait. If the two alleles were different (one wrinkled and one smooth), he called it *heterozygous.*

Within a decade of Mendel's rediscovery, scientists across the globe had accepted his laws as the basis of heredity in most plants and animals. This knowledge could be put to practical use; in the United States, large agricultural stations were set up to use Mendelian principles to create better corn, beans, wheat, tobacco, and other crops.

FIRST ATTEMPTS TO LINK GENETICS AND EVOLUTION

While today's scientists know that Mendel's ideas are entirely compatible with evolution, this was not so clear at the beginning of the 20th century. Researchers believed that the concept of genes might explain variety within a species, but it did not say anything about how one plant evolved into an entirely new type. This had to involve more than just remixing the genes that already existed in a population; otherwise, a new species might simply revert to its previous form (for example, Galápagos finches might give birth to chicks with the beaks of their ancestors). Besides, if the chromosome theory was correct, genes lay on the chromosomes. Flies had only four pairs; humans had 23. The extras must have come from somewhere.

Bateson thought he saw an answer in what he called *discontinuities:* cases where individual plants and animals underwent strange duplications of their parts. An insect might have extra body segments, giving it additional legs; a goat might grow too many horns. Such dramatic mistakes, Bateson felt, were the kind of jumps that could produce new species. But these changes were sudden and dramatic, whereas Darwin had claimed that evolution was a very gradual process during which traits change in very small ways over long periods of time. Were discontinuities the fuel of evolution? Scientists were divided on the issue. Traditionalists, known as "biometricians," were uncomfortable with the either-or nature of genes; they dismissed discontinuities as monsters that would be quickly weeded out by natural selection.

Biometrics is mostly concerned with organisms as wholes, measuring them and plotting their physical traits on charts. Biometricians had a chart for size, for example: a curve with most members of a species clustered around the middle, near the average. (Unusually small individuals sloped off to the left, and extraordinarily large ones tapered off to the right.) There might be another chart for speed, hearing, and so on.

Biometricians expected to see natural selection at work at the edges of the curve. Under normal circumstances, the average specimen was probably the most fit. But if the environment changed, having a body that was slightly larger or smaller than the norm might be advantageous. The trait would be favored— it would undergo "positive selection," and within several generations the curve would be pushed toward a new norm. As this happened for many traits, new species would arise.

Geneticists believed that fundamental changes in the hereditary material were more important. In 1901 de Vries had already proposed a mechanism by which this might happen: sudden, discontinuous changes which he called *mutations.* The problem was trying to tell the difference between these changes and the normal variation found within a species. Finding a solution would require studying an organism very closely in the lab.

THE QUEST FOR MUTATIONS: THOMAS MORGAN'S FLY LAB

Thomas H. Morgan (1866–1945) made great steps forward in understanding the nature of genes and mutations through his work with fruit flies. He brought *Drosophila melanogaster* into the research laboratory because the insect took up very little space (thousands could be kept alive in a glass jar, and kept happy on a diet of mashed bananas), and it reached sexual maturity just two weeks after birth. This led to a very fast rate of reproduction that was ideal for genetic studies. Additionally, the small number of chromosomes in the fly would simplify things as Morgan looked for the physical locations of genes.

Morgan's lab developed very efficient methods of *screening* flies: observing insects closely under the microscope, by the thousands, in hopes of finding unusual characteristics. The work paid off with the discovery of dozens of mutations within the first few years of work. Mutations followed a pattern that was easy to recognize: Traits that had not appeared at all in previous generations suddenly appeared and were passed along as dominant or recessive genes. By crossing mutants with other flies, the laboratory identified genes responsible for the color of the fly eye and body, its size, and the shape and structure of wings. By 1915, the researchers had identified 85 mutations and thus 85 genes.

This was an impressive number, but it clearly fell far short of the total number of genes in the fly. The search for new ones was slow going, because the only way to obtain mutations was to wait for nature to produce them. Morgan had tried various things to try to speed things up; one of his students succeeded. After leaving the lab Hermann Muller began exposing flies to radiation. By 1927 Muller had produced hundreds of new mutations in a new laboratory he established in Texas. The discovery earned him the 1946 Nobel Prize in physiology or medicine.

Thomas Hunt Morgan at work in his lab *(A. F. Huettner/the Archives, California Institute of Technology)*

Morgan did not particularly care what genes were made of—the question was still completely open. Even so his lab and other groups were making progress by proving the chromosome hypothesis. One important discovery was the reason that flies and other animals came in two sexes. One of the chromosome pairs in females consisted of two identical X chromosomes, whereas males had one X and one smaller Y. When Morgan's lab discovered specific genes located on these chromosomes, it proved that genes were single physical units, rather than abstract objects that might be scattered across many chromosomes. Morgan's group used these findings to create the first chromosome map that marked the positions of specific genes. By 1915, this had evolved into a model where genes were seen as "beads on a string" of the chromosome.

The shapes of chromosomes vary slightly from species to species of fruit fly. Just as comparing the shapes of teeth helped paleontologists place fossils into their proper place in the evolutionary tree, the forms of chromosomes said something about their history. The scientists began drawing evolutionary trees of the fly based on these characteristics. Evolution had now dipped below the large, visible features of plants and animals to the level of the cell. In a few decades it would go even deeper, directly reading the genetic code—the language in which evolution is written.

UNDERSTANDING POPULATIONS THROUGH MATHEMATICS

Scientists still did not know how important mutations and natural selection were to evolution. Some geneticists boldly claimed that new species could arise from accumulations of mutations alone, in a process they called *genetic drift.* This would be a "gentler" form of evolution in which natural selection played a much smaller role, appealing to those who did not like the view of nature as a place of constant struggle, in which every organism had to pass a fitness test. Biometricians, on the other hand, still insisted that natural selection was the main force behind evolution. They thought that mutations were aberrations like Bateson's monsters, which would usually be eliminated very quickly. Even if an organism with a mutation survived, it was not clear that the trait would survive and spread enough to create new species.

Mathematicians began trying out models of how genes might move through populations. George Udny Yule (1871–1951), a Scottish statistician, predicted that dominant genes would multiply in a population and wipe out recessive traits. The American geneticist William Castle (1867–1962) calculated that with no natural selection at all, the frequency of genes in a population would remain stable over many generations.

They were creating a new field—*population genetics*—and had to face the problem that while mathematics and laws are clean, nature is disorderly. Even the transmission of single Mendelian traits like smoothness or wrinkledness in peas was difficult to understand. What seemed precise and systematic in the laboratory became chaotic in the real world, where there were no breeders to precisely control which plant fertilized another or which flies mated. Population genetics had to cope with the disorder and consider anything that might influence patterns of heredity for genes.

Darwin had predicted that natural selection would work hardest when the pressure on a population was extreme—when a high number of predators lurked in the neighborhood, or when there were far more mouths than the food supply could possibly feed, or when environmental conditions changed. There would be pressure all the time on animals such as birds—he calculated that in just 200 years, eight pairs of swifts could produce 10 thousand billion billion billion descendants if nothing kept them under control. But what about animals that were much less fertile? Darwin used one of the slowest-breeding creatures on Earth, the elephant, to do some calculations. If the animals bred between the ages of 30 and 90, and had only six offspring, a single pair would have nearly 19 million descendants after only 750 years.

Thus natural selection would be an issue even for elephants; the question was how to detect and measure it. The mathematicians Godfrey Hardy (1877–1947) and Wilhelm Weinberg (1862–1937) independently came up with an answer based on statistics and probability. They arrived at the description of the behavior of genes, and today their conclusions are known as the *Hardy-Weinberg law.* In humans, other animals, and plants that reproduce sexually, the formula consists of the following steps:

1. Determine the frequency of genes among the adults
2. Find out which types of adults mate with each other

3. Estimate the frequency of genes among their offspring using Mendel's ratios
4. Discover how many of the offspring survive to reproduce

The formula could be used to test hypotheses about populations and evolution; for example, it verified William Castle's prediction that in a large population with random mating and no natural selection, the genetic makeup would stay the same over many generations.

The Hardy-Weinberg law can be simulated using a bag that contains an equal number of red and black checkers. These represent alleles for color, and the goal is to track whether one color becomes more frequent over time. A person reaches in and pulls out a pair—this is like an organism, which has two copies of a certain gene. The result might be two blacks, two reds, or one of each color. To obtain a whole population, two more checkers are removed, over and over, leaving some checkers in the bag (these represent organisms that do not reproduce in one generation). At the end of the experiment, about a fourth should be pairs with two reds; another fourth should be two blacks; the rest (half) should be mixed pairs containing one black and one red checker. This ratio of ¼:½:¼ is what the Hardy-Weinberg equilibrium predicts for cases of genes with two alleles.

If there are not equal numbers of alleles to begin with (if one starts off with more black checkers than red), the formula has to be adjusted, but it still shows that the balance will stay nearly the same over many generations. Another adjustment has to be made when a population has three alleles for one gene (as if peas could be round, wrinkled, or hairy). This alters the formula and the predictions, and it can be simulated by putting three colors of checkers into the bag, still drawing only two at a time (because a single organism comes with two alleles for each gene).

If the results of the experiments do not confirm Hardy-Weinberg's predictions, something is probably influencing the process by which checkers are selected. Red checkers might be heavier and collect at the bottom of the bag. A person who reaches in and selects checkers from the top will pull out more black ones; someone reaching deeper will get more red—like natural selection. If the checkers are magnetic, certain pairs will attract each other and become more likely to be drawn out together (as if mates are not randomly choosing each other). Any of these factors could change the ratio of "colors" in the gene pool over time.

Scientists need genetic population models to understand how disease organisms develop resistance to drugs and in attempts to save endangered species. The formula shows that mating patterns are important and that large populations behave differently than small ones. A small population has problems because there is more mating between close relatives. This allows dangerous mutations to spread fairly quickly, and it is more likely that individuals will inherit two copies of a harmful allele. This is an important reason why endangered species should be protected: They are threatened not only by predators, but also by the dangers that arise from inbreeding.

An example can currently be seen in the Tasmanian devil, a species of marsupial that inhabits the island to the southeast of Australia. The devils survived a bottleneck recently in history, so they nearly all descend from a small group of animals. They have experienced such heavy inbreeding that their immune systems can no longer tell the difference between "self" and "not self." This is causing a spread of cancer that has wiped out nearly half of the species. The animals engage in fierce fights in which they bite each other on the face. When that happens, tumor cells move from one devil to the next. In most species such foreign cells are rejected by the immune system, but the Tasmanian devils are so closely related that this does not happen.

THE BEGINNINGS OF UNIFICATION: RONALD FISHER AND SEWALL WRIGHT

The Hardy-Weinberg formula described how a species' genes would be reshuffled from generation to generation, and it provided a way to measure how much selective pressure was operating on a particular gene. But it still did not address the big question of whether evolution was driven by mutations. That problem would be tackled by Ronald Fisher (1890–1962), a talented mathematician who developed an interest in evolution at an early age. (He also had a sharp wit, once saying, "To call in the statistician after the experiment is done may be no more than asking him to perform a postmortem examination: He may be able to say what the experiment died of.") As a university student he became convinced that the dispute between the geneticists and the biometricians was the result of a misunderstanding, and he used mathematics to prove that there was no real contradiction between the two positions.

He started by redefining the problem. Geneticists were right that mutations were obviously a source of new traits in a species. But once they existed, what happened to them? The first step toward making a new species was for new forms of genes to become part of the mainstream. Evolution could not be driven mostly by mutations, Fisher showed, because they would have to be happening in huge numbers, all the time. Instead, if a mutation happened and was transmitted through a population, it became part of the species' gene pool and behaved like any other gene. At the beginning the allele would be rare, but it would spread even if it only offered a tiny advantage in reproduction. Mixing the old and new features would create exactly the small-scale, continual variation that natural selection worked on.

Selection was poorly understood, Fisher believed, and he wanted to be sure that Darwin's concept was possible—in other words, that natural selection could change the genetic profile of the population and generate new species. To settle this question mathematically he needed numerical values, so he transformed the idea of variation into variance—a quantity. In his 1930 book *The Genetical Theory of Natural Selection,* he also introduced a quantitative concept of *fitness,* meaning the degree to which a species is adapted (or not) to its environment. Most species were already fit, he suggested, having been produced by natural selection. However, their current state was the result of the environment of the past, rather than the one they confronted now. This meant, for example, that human nature was mostly defined by pressures of natural selection early in the history of the species, when humans lived as hunter-gatherers, and much less by the modern technological age.

Fisher was strongly influenced by physics and drew comparisons between physical principles and those at work in biology. Genes in a population, he wrote, were like atoms in a cloud of gas: The behavior of a cloud could be described by statistical formulas, as probabilities, even though it was impossible to predict how a single atom might behave at a given time. Similarly, an individual might be an almost random collection of alleles, but a whole population could show patterns, the way gravity, temperature, and other factors affect the behavior of a gas.

Another question was whether Bateson's dramatic mutations would make evolution go faster than smaller ones. Fisher used the analogy of a microscope to show that big changes are usually harmful. Organisms are usually "in focus," because natural selection has tuned them to their environments. When looking at something under the microscope, a big movement of the lens is likely to turn everything into a blur, just the way a huge mutation is likely to seriously disrupt the relationship between an organism and the environment. The most successful mutations were likely to be almost insignificant, Fisher wrote—like very slight adjustments in the focus.

In 1919, Fisher obtained a job at an agricultural station north of London, which had been collecting a huge amount of data on breeding. He became the station statistician and began genetic studies of poultry, snails, and mice. This work eventually led to *The Genetical Theory of Natural Selection,* in which he formulated what he called the Fundamental Theorem of Natural Selection: "The rate of increase in fitness of any organism at any time is equal to its genetic variance in fitness at that time." It is clear that he hoped to understand how variety influences evolution, but what exactly he meant has been debated ever since. Perhaps the answer lies once again in physics: the more elements there are in a gas, the more possibilities there are for making new substances, and the more chances there are that one of them will be "fit" when conditions change. But if all the members of a species are nearly identical, selection will work very slowly, and it will take a species much longer to become fit.

Fisher had another great insight. Anything could be subject to natural selection: "languages, religions, habits, and customs, rocks, beliefs, chemical elements, nations, and everything else to which the terms stable and unstable can be applied." The idea made Fisher start thinking about humans, and he devoted several chapters of *The Genetical Theory* to thoughts about eugenics (discussed in the previous chapter). He founded the Cambridge University Eugenics Society, which concerned itself with the fact that lower socioeconomic classes were having more children than the upper classes and promoted a "voluntary sterilization" program, but the society's proposals were consistently voted down in the British Parliament.

Fisher's fundamental theorem led Sewall Wright (1889–1988) to think of fitness as a sort of rugged *fitness landscape,* full of peaks that represented individual species. Single plants and animals showed various degrees of fitness—the most fit falling in the middle of the curve, at the peak. Species would move toward fitness because genes or traits on the edges of the curves would usually be eliminated by natural selection.

Wright introduced the concept of genetic drift, which meant that the genes of a population would inevitably change, even with no great pressure from selection or the environment. Random events would keep generating new variations, and over time their frequency would slowly change in the population. For Wright, accidents and pure chance played a more important role in evolution than Darwin had thought. However, he still saw natural selection as a powerful force, whittling on the peaks, moving the species as a whole toward peaks of fitness.

Wright's laboratory work involved studies of the colors of guinea pigs, using animals that had been inbred and crossbred for many generations. William Castle, his professor at Harvard, believed that color was controlled by a key gene, but if so it ought to follow Mendelian rules. Wright's experiments proved that several genes were involved in producing color, probably by working together. Today we know why: Genes contain the

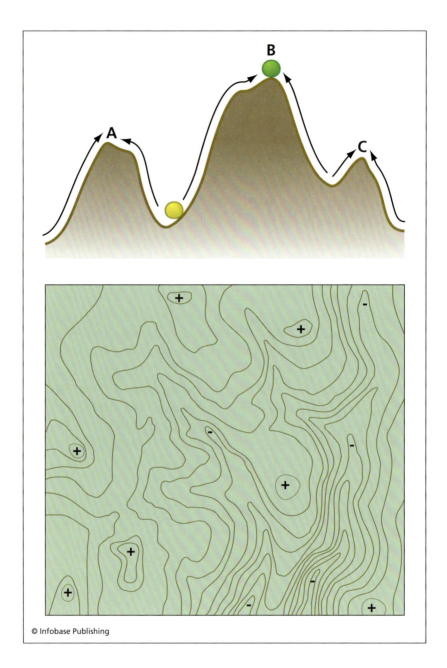

recipes for molecules that cause chemical and structural changes in cells. Everything that happens—from the simplest chemical reactions to the construction of complex structures like a

(*opposite page*) Sewall Wright's "fitness landscape" is like a topographic map, comparing an organism's degree of fitness in its environment to a landscape with peaks and valleys. Over time, populations as a whole move toward peaks and out of valleys.

wing—requires a mutual effort. Several genes might collaborate to create one feature, and ultimately, different mutations might have the same effect on an organism.

Wright also wondered about the problem of distinguishing between genetic and environmental influences on an organism. He developed a method called *path analysis* that examines complex sets of relationships to try to link causes and effects. Today this method is widely used in the social sciences to determine the environmental factors that influence behavior.

One result of Wright's work was a careful mathematical description of how mating systems influenced heredity. His formulas allowed scientists to calculate the number of generations it would take for genes that were being selected at various "strengths" to spread through the population. These studies, along with the formulas of Fisher, became the basis of scientific animal breeding in the United States and elsewhere.

RECIPES FOR ISOLATION: J. B. S. HALDANE

The Englishman John Burdon Sanderson Haldane (1892–1964) was one of the most colorful figures of early 20th-century science. For example, during World War I he built bombs in the trenches on the front lines in France, launching them fearlessly (many said recklessly) at the enemy. Once he mounted his bicycle and rode it in plain view between trenches, having calculated that the Germans would need too long to react to shoot him. Later he subjected himself to dangerous experiments concerning the effects of gases and diving on the body. One experiment with oxygen caused seizures that crushed some of his vertebrae. Haldane brushed these off as the sacrifices that had to be made in the name of science. When trials in a decompression chamber perforated his eardrums, he wrote, "The drum generally heals up; and if a hole remains in it, although one is somewhat deaf, one can blow tobacco smoke out of the ear . . . which is a social accomplishment."

Like Fisher, Haldane thought that some of the major questions about evolution would have to be solved through mathematics and population studies, since they had happened in the past and could not be reproduced through experiments. Natural selection not only had to explain how species changed, it also had to show that evolution could have occurred within the amount of time provided by the age of the Earth. In 1924 Haldane began a project called "A Mathematical Theory of Natural and Artificial Selection." One question he posed was whether scientists could ever witness natural selection in the act, because it happened slowly and gradually.

The answer depended on how natural selection worked in a living species. If one knew how frequently a trait appeared in a population and then studied how it rose and fell over generations, Haldane wrote, it ought to be possible to measure how strongly natural selection was acting on the trait. Even if it offered only a very small advantage, as little as one-tenth of one percent, with enough time it could become frequent in a population. He plotted what might happen after thousands of generations. By comparing predictions to real measurements of the frequency of alleles in a population, one ought to be able to detect how much of an advantage any particular mutation provided.

Haldane extended his calculations to unusual hypothetical cases, such as the following: Suppose that two recessive genes that affected eyesight were circulating in a population. On its own, each led to poor eyesight, but if someone inherited both genes, there would be an improvement. Haldane could predict whether the trait would survive and how frequent it might become. Bizarre behavior such as cannibalism in fish did not seem like it could be influenced by genes. If parent fish ate their offspring, how could "cannibalism genes" survive natural selection? Haldane proved that it could. In a population of a few million fish, the first cannibals would be more likely to eat the off-

Peppered Moths and Finches

The case of the peppered moth seemed to be a perfect example on which to test Haldane's ideas. This moth, found in the British islands, was normally a speckled grey color, but in the mid-1800s collectors had begun to find specimens near Manchester that were completely black. By the end of the century, nearly all the moths in the area were black. With the arrival of the industrial revolution, roofs and chimneys were blackened by soot, and the dark color probably made the moths completely invisible to the birds that preyed on them. This looked like natural selection in action, and Haldane began counting and calculating the advantage that blackness conferred on the moths—it seemed enormous, in the range of 30 to 50 percent.

Populations of speckled moths living in heavily industrial regions usually darkened, whereas insects in the countryside remained lighter and speckled. When clean-air laws began to be passed, the population lightened again. But nature is not simple, and there were exceptions to the rule; sometimes populations of black moths could be found in the countryside. Perhaps they were migrating, or perhaps birds were not the insect's only predators.

Other cases of the selection of features were harder to explain. In the 1940s, a young scientist named David Lack (1910–73) traveled to the Galápagos to try to find out why the archipelago should host thirteen species of finches rather than just one. The finches had been a key discovery for Darwin during his visit to the islands. Lack originally believed that differences in the birds' beaks were simply caused by mutations and drift. Over time, however, he began to believe that natural selection had played a strong role. The beaks gave the species different

(continues)

(continued)

lifestyles, which meant they did not compete for food. One type of beak allowed the bird to forage like a woodpecker, digging at insects under the bark of trees. Another species could capture and eat insects in flight, and yet another had become a sort of vampire, pecking at the veins of other birds and drinking drops of blood. While species living on different islands (isolated through geography) might be similar, species that shared an island were usually quite different; to coexist they had isolated themselves by moving into a different niche. This would submit them to different evolutionary pressures and result in the selection of different features. The work convinced Lack that even slight, seemingly "meaningless" differences between the birds were probably the result of selection, rather than drift alone.

spring of other fish than their own. In fact, at first they would have an advantage over noncannibals because they would have more food and better chances for survival. However, in the long term, or in a small population, the gene would be harmful and unlikely to spread.

The "Mathematical Theory" reminded readers that isolation was an important part of the recipe for promoting the movement of recessive genes and creating new species. The obvious case was when a geographical barrier separated members of a species, but Haldane discussed four other types. *Psychological isolation* referred to behavior and sexual preferences—the tendency of animals (and people) to prefer certain types of mates. Plants, too, could become "psychologically isolated" if acquiring a new color or scent attracted different types of insects. An example of *anatomical isolation* is that organisms of a similar size tend to mate. *Temporal isolation* would occur if a mutation changed an organism's breeding or flowering time. Finally, *selective fertiliza-*

tion could restrict which other members of its species were pollenized by a particular plant.

THE RISE OF NEW SPECIES: THEODOSIUS DOBZHANSKY AND ERNST MAYR

Fisher and Haldane had begun to bring genetics and evolution back together. Theodosius Dobzhansky (1900–75) played a crucial role in further linking the two fields. Born in Russia, Dobzhansky immigrated to the United States in the late 1920s and began working in the laboratory of Thomas Hunt Morgan. Dobzhansky hoped to find genes that might act as a snapshot of how varieties were becoming different species. To do so he would have to go outside the laboratory, where inbreeding had produced flies with nearly identical genes.

Dobzhansky collected specimens of flies over a huge area, from Mexico to Canada. One might think that groups of flies living near each other would stem from the same ancestor, and thus be very similar, but in each of the places he looked he found a surprising amount of variety. Single populations contained different genes, even different versions of the same chromosome. It was not possible to define a group on the basis of a particular set of chromosomes; instead, Dobzhansky had to describe it with statistics, describing which versions of the chromosome were most common. Another group of flies living nearby might have a much different set.

What kept groups separate, Dobzhansky realized, was sex. Anything that stopped groups from mating with each other was the first step toward evolution. If a small group became isolated, it probably would not have exactly the same proportion of genes as the large population it came from; they would get shuffled around in different proportions and respond to natural selection in their new environment. Ernst Mayr (1904–2005) called this the *founder effect.*

Dobzhansky and Mayr believed that some mutations would be harmful, of course, but others might be neutral; in other

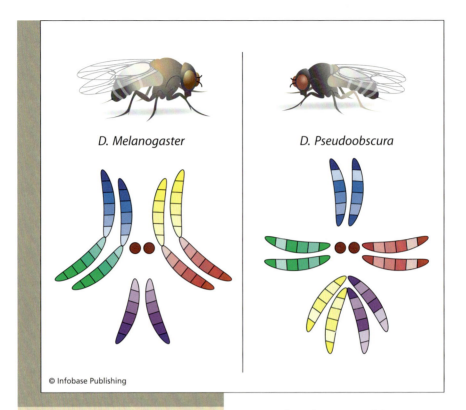

D. Melanogaster   *D. Pseudoobscura*

© Infobase Publishing

Microscope studies of the shapes and arrangements of chromosomes in different species of fruit flies gave Theodosius Dobzhansky unique insights into fly populations and their regional habitats.

words, they would not immediately have an effect on organisms. But they would increase the overall genetic differences between two populations. Eventually this would cause them to become genetically isolated—no longer able to mate. (If they attempted to do so, they would produce sterile offspring, as mating horses and donkeys does.) In his 1942 book, *Systematics and the Origins of Species,* Mayr wrote, "A new species develops if a population which has become geographically isolated from its parental species acquires during this period of isolation characters which promote or guarantee reproductive isolation when the external barriers break down."

This version of evolution satisfied both the geneticists and the Darwinians, and it finally brought the two approaches together. "Nothing in biology makes sense except in light of evolution," Dobzhansky wrote, and he had found a way to make sense of a great deal, including what Ernst Mayr was observing as he studied birds in New Guinea.

Mayr was following in the footsteps of Darwin, Alfred Wallace, and other naturalists who had wandered the world looking for new species and studying their habits. He spent many years observing and classifying the "birds of paradise" that had helped Wallace establish his career. Like many classifiers, he faced the problem of having to decide which birds constituted separate species and which were subspecies—groups within a species that had developed particular characteristics but were not yet truly isolated.

Mayr noticed that in New Guinea, many subspecies had specific ranges of habitats and particular characteristics—their tails might be shorter or longer, pointed or square. They continued to mate with each other, but they preferred different regions. This suggested that they might become easily isolated. For the types to become true species, the separation did not need to be permanent; it just had to last long enough for significant differences to develop between populations.

EVOLUTION AS A GAME: JOHN MAYNARD SMITH

John Maynard Smith (1920–2004) left a career in engineering and mathematics to become a devoted student and eventually a close friend of J. B. S. Haldane. One of his first projects was to study the effects of inbreeding in flies in Haldane's lab. Inbred flies could not cope with high temperatures very well, and Maynard Smith thought this could be used for another test of Lamarckianism, which was undergoing a small revival in Russia.

The Communists believed that human nature was dictated by the economic system in which they were raised. So the new economic system put into place after the Russian revolution ought to fundamentally change human nature, possibly even people's genes. State research programs were set up to

John Maynard Smith (*Colin Atherton, University of Essex*)

investigate whether Lamarck might not have been right after all: Could an organism's lifetime experiences (for example, living under a Communist state) somehow become part of its genetic baggage? DNA's role in heredity was still unknown, and scientists were still puzzled about the pace and mechanisms of evolution. After six months of work Maynard Smith could not discover any evidence of Lamarckian inheritance in flies.

He next became interested in the phenomenon of aging, which seemed to contradict evolutionary principles. Logically, natural selection ought to favor creatures that live to a very old age, because they would have more offspring than animals that died at a younger age. But in the wild, he realized, organisms rarely lived to advanced ages, so natural selection would not have a chance to favor them. It was interesting, though, that various parts of the body tended to break down at about the same time. Natural selection could certainly improve one weak part so that it kept up with the rest.

Maynard Smith spent several years unraveling the ways that fruit flies chose their mates, demonstrating that the males of some species have to perform an elaborate courtship dance to attract females. The ability to do the dance is based on genes. Human behavior was obviously permitted by human genes— but to what extent was it controlled by them? This was a new perspective on psychology that would become a major theme of the second half of the 20th century.

Another theme that interested Maynard Smith, Haldane, and a young student named Bill Hamilton (1936–2000) was altruism. Could natural selection ever favor an organism which deliberately sacrificed itself for other members of its species? Clearly saving one's own child would promote the survival of family genes into the future, but saving others might have the opposite effect—favoring foreign genes, rather than one's own. A curious alternative had been proposed: selection might not always act on individuals alone; it might work on groups, or whole species. Maynard Smith was skeptical. With mathematics he proved that this could happen only if the payoff was incredibly high. Otherwise, natural selection always acted on individuals. It could select individuals who made sacrifices for their offspring, but this was completely in line with traditional Darwinian natural selection. At the level of genes, altruism was just another type of hereditary selfishness.

Maynard Smith's most important contribution to evolutionary theory came from considering other types of behavior that did not seem to fit classical natural selection. When males of many species fought for females, they usually did not cause serious or fatal damage. Snakes often tangled with each other but did not bite (venom can kill snakes too), and deer bucks stopped short of goring each other. It was not clear why an animal should hold back: Killing a competitor would probably increase an animal's chances of putting more of its genes into the next generation.

Maynard Smith achieved a breakthrough when he learned of "game theory," a mathematical approach that was being used in economics to study the behavior of groups. Here, too, scientists were asking questions about selfishness and altruism, but phrased in a slightly different way. To win a game, for example, it might seem that the best strategy would be to fight as hard as possible and do a great deal of damage to the other players. But this did not consider that games often take place in several rounds, and bad behavior might provoke severe reprisals the next time. In the marketplace, buyers and sellers have to make decisions based not only on the outcome of a single transaction, but they also have to think about the effect that their behavior has on other people's decisions. Competition for mates works that way, Maynard Smith thought: If an animal seriously wounds its competitors, it might open itself up to brutal acts of revenge. Thus a more restrained type of struggle that stopped before things got out of hand might not be altruistic; it might simply be the best strategy for an individual's long-term survival and reproductive success.

The question was whether this could be demonstrated mathematically. Maynard Smith studied game theory in the mathematical work of John von Neumann, Oskar Morgenstern, and an American named George Price (1922–75). Maynard Smith and Price started to use mathematics to analyze the effects that different competitive tactics would have on evolution. They performed simulations using five character types—from a "hawk" that always behaved aggressively to a timid "mouse." The goal was to find an *evolutionarily stable strategy,* or ESS, that would permit the population to survive in a relatively constant state. They discovered that a population consisting only of hawks would not last; likewise, if nearly everyone was a mouse, one aggressive mutant could come in and disrupt everything. The most stable group was a mixed one in which organisms copied each other's behavior, and in which retaliation for aggressive acts was permitted.

One use of the new approach was to explain territorial behavior. Instead of directly competing for mates, butterflies and other species may protect the territory in which they will mate. Their behavior falls into an ESS pattern in which the "owner" of a territory acts aggressively and the intruder gives way; other strategies

© Infobase Publishing

are unstable because they result in constant competition. Several years later, scientists discovered a species of California lizard that followed the rules of Maynard Smith to the letter. Game theory was not only an abstract way of thinking about evolution on the chalkboard; it also described nature.

A type of mathematics called game theory allowed John Maynard Smith and George Price to make sense of behavior such as competition for mates and territorialism, which seemed difficult to fit into evolutionary theory. Their famous example of "hawk" and "dovelike" behavior demonstrated, for example, assigned numerical values to the different "payoffs" that could be obtained when the competitors behaved in different ways.

Toward the end of his life Maynard Smith focused on the question of why sex evolved and how it might give organisms an advantage. The earliest single-celled organisms reproduced without sex, simply by dividing. This seemed like it would push evolution quickly, because it would eliminate the problems of finding mates and the possibility that they might carry defective genes. But was it really better than a sexual mode of reproduction? Some organisms—from particular types of bacteria to blackberries—can reproduce in either way, through sex or by *cloning* themselves. Maynard Smith showed that in the short term, asexual reproduction might be an advantage, but over the long term it would likely lead to deadly accumulations of mutations. Sex provided a method of clearing out mistakes by mixing up the genetic material, and also as a backup system. If an organism inherited a copy of each gene from two parents, it would have a spare in case one of them got damaged.

"PUNCTUATED EQUILIBRIUM": STEVEN JAY GOULD AND NILES ELDREDGE

Evolution should explain both life on Earth today and its history, and fossils have been the starting point for some controversies within the theory. So few fossils have survived that they provide only a tiny glimpse into the past; even so, if evolution was a slow, steady process, as Darwin believed, they ought to reflect that. Yet there seemed to have been epochs, hundreds of millions of years ago, in which "bursts" of evolution took place. To account for this, in 1972 Stephen Jay Gould (1941–2002) and Niles Eldredge (1943–), two American paleontologists, proposed a concept called *punctuated equilibrium.*

Eldredge had been studying *trilobites,* fossil animals that thrived throughout the oceans of the world over an extremely long time, from about 540 million to 245 million years ago, and then disappeared. He expected to find that individual species had undergone gradual changes over long periods. Instead, some species seemed to have changed very little at all. Gould was discovering the same phenomenon in his work on snails; some fossil species are nearly identical to those living today. When he was invited to contribute a paper to a book about evolution and the fossil record, he invited Eldredge to write it with him. In it they introduced the concept of punctuated equilibrium.

The idea is related to Sewall Wright's "fitness landscapes," which suggested that if the environment stays constant over long periods of time, species might, too. Once an organism had been "optimized" for its environment, changes would almost always be harmful. So the main function of natural selection would be to weed out unusual variants and keep the species stable.

Gould and Eldredge said that after long periods of little change, sometimes there are sudden evolutionary bursts in which many new species arise. In fact, the development of new species might be the exception rather than the rule. It might be some-

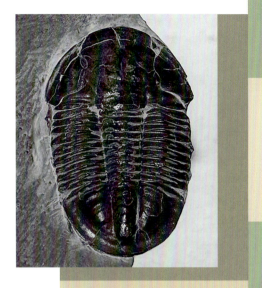

Trilobite fossil. The species is *Asaphiscus wheeleri*. These come from the Cambrian-age Wheeler Shale from Millard County, Utah, near Antelope Springs. The fact that trilobites survived almost unchanged for 250 million years while other species changed much more quickly was one factor that led Stephen Jay Gould and Niles Eldredge to propose the hypothesis of "punctuated equilibrium." *(Wikipedia)*

thing that happened very quickly, possibly within the space of just tens of thousands of years, after millions of years of quiet. They cited the *Cambrian explosion,* which occurred from 540 to 500 million years ago, as an example. This period saw the origin of nearly all of the basic animal forms known today. There is a well-preserved record of their existence because for the first time, animals had hard body parts which survived long enough to be fossilized.

To support the concept of punctuated equilibrium, Gould and Eldredge also had to think a bit differently about selection itself and its relationship to extinctions. If natural selection always acts on individuals, then why do some species as a whole survive, and others die out? This led them to think that sometimes selection might work at a different level, in a process called species sorting. Some environments might encourage the spin-off of new species; others might prohibit it. A limited number of birds might develop in the Galápagos, for example, because of the islands' limited resources. The fact that different species of birds often eat similar foods and are hunted by the same predators might lead to a form of competition between different species, as well as within them. The South American mainland could permit many more spin-offs—many more species of birds—because of its larger territory and richer resources.

More traditional Darwinists have objected to this idea because by definition, natural selection is something that works within species, rather than between them. Darwin introduced the concept to explain why today's species exist and why they have certain features. Some have claimed that there was nothing special about the Cambrian period; what looks like an explosion is simply due to missing information, the usual gap in the fossil record.

In a way, Gould, Eldredge, and their critics may all be right. The Cambrian explosion may have produced a wide range of new types of animal bodies without any special "rules" at work in evolution. It does not necessarily take many mutations to produce organisms with completely different shapes and forms. Without prior knowledge, not many people would guess that a neuron (which looks like a tree with complex networks of roots and branches) and a doughnut-shaped red blood cell were produced by exactly the same set of genes.

Currently laboratory experiments have shown how small changes in an organism can have large effects. When Martin Lipp's group at the Max Delbrück Center in Berlin shut down a single mouse gene called CXCR5, mice failed to develop lymph nodes. Normally this would destroy their immune systems. But their bodies compensated by building smaller, bulblike structures in other places in the lymph system that provided almost normal immune functions. This points out a weakness in the philosophy of "intelligent design": sometimes many different designs can work equally well. There is not necessarily one that is better designed than the others—unless it proves better at surviving the challenges of natural selection.

The analysis of the genome of maize has shown that it only took changes in five genes to produce today's huge ears of corn from the tiny cobs cultivated by ancient Central Americans. Another example came in 2004 when Nadia Rosenthal, who wrote the foreword to this book, activated a single gene in an unusual way in a mouse at the European Molecular Biology Laboratory in Italy. By making muscle tissue produce a protein called IGF-1, the researchers obtained an animal that lived much longer than normal. It had such huge muscles that her lab nicknamed it the "Schwarzenegger mouse."

It will probably never be possible to directly measure how many new genes were needed to cause the Cambrian explosion, because the DNA in fossils older than about 100,000 years has deteriorated too far for analysis. Some information may come from comparing genes of the descendants of Cambrian animal types. This type of study is discussed in the next chapter.

There may also be other explanations for the Cambrian explosion, having to do with mechanisms in the genome that scientists are only now beginning to understand. One of the startling results of the analysis of the complete human genome sequence has been the discovery of millions of *transposons,* small DNA sequences that can copy and paste themselves into new positions in chromosomes. Plant and animal genomes are full of these sequences, which have replicated and spread through DNA-like viruses. As they jump around, they move to random places and often disrupt genes. The evolution of new transposons might have caused species to change much more rapidly than usual.

Another concept developed by Gould, in an article with Richard Lewontin in 1979, was the notion of *spandrels.* This was a metaphor based on the architecture of buildings like the Cathedral of San Marco in Venice, which has a huge dome. The builders supported the dome by placing it atop two arches that intersect each other in the form of an X. This cuts the dome

Stephen Jay Gould (*Columbia University*)

into four curved triangles, called spandrels, which have been decorated with intricate mosaics. The spandrels themselves have no function; they are simply there because a dome is mounted on two arches. The same thing happens in evolution, Gould and Lewontin wrote. Some features of an organism are not really selected; they are just along for the ride, the by-products of structural features. Once they exist, they can become useful if natural selection goes to work on them.

Gould believed that many aspects of human mental behavior are spandrels. People can read and write and make music, for example, because their brains have a certain structure. But these skills evolved recently in our species history, and probably have not played a role in natural selection. In contrast, the ability to hunt, to solve problems, and to speak probably did help in the survival and reproduction of early hominids. As these skills were selected, they had wonderful by-products: the ability to write and create works of art.

Other Darwinist thinkers, such as John Maynard Smith, have been uncomfortable with the idea of spandrels. Ernst Mayr and the bird-watching community had shown over and over again that traits which seem unimportant actually have functions that could easily make them subject to natural selection. On the other hand, no one really believes that everything is under the scrutiny of selection, in every species, all the time. Our bodies preserve ancient features that were important millions of years ago. Mutations have not deleted those features because they still have functions and organisms suffer if they are lost.

How much influence spandrels have on evolution is partly a question of the point of view. Selection works on whole organisms all the time, from birth until the end of their reproductive stage—even longer, in species in which parents need to stay around to take care of their offspring. Because all phases of development are crucial to survival, selection can zoom in on any feature at any time it is active. From a functional point of view, many genes act like "spandrels" at some point in an organism's life; at others they are hugely important structural elements, needed to hold up the entire organism.

Gould became one of the most widely read authors on evolution, writing articles for the popular press and dozens of excellent books. However, many evolutionary biologists criticized his ideas, mostly on the grounds that molecular biology and evolution had not yet truly become united. The debate surrounding Gould's work echoes the much older debate between the biometricians and the geneticists. Chapter 6 describes some of the ways scientists can now, finally, address these issues.

EVOLUTION FROM THE GENE'S POINT OF VIEW: RICHARD DAWKINS

In 1976 Richard Dawkins wrote a book called *The Selfish Gene* which took an extreme view of natural selection: Its real effect was to promote or hinder the survival of particular forms of individual genes. If having a cleft chin is the result of one gene, and this feature becomes advantageous for some reason, then the result is a rise in the frequency of the cleft-chin allele in a population. Dawkins sees evolutionary competition ultimately as a struggle between single alleles, doing what they can to ensure a ride on chromosomes by pushing other variants aside.

Taking this point of view to the extreme, organisms as a whole can be regarded as "simply" vehicles for reproducing genes. Far back in history life began as a self-replicating molecule, then it became a collection of ever more complex molecules. Dawkins wrote, "Now they swarm in huge colonies, safe inside gigantic lumbering robots, sealed off from the outside world, communicating with it by tortuous indirect routes, manipulating it by remote control. They are in you and me;

they created us, body and mind; and their preservation is the ultimate rationale for our existence. They have come a long way, those replicators. Now they go by the name of genes, and we are their survival machines."

Dawkins calls the behavior of alleles "selfish," by which he usually means something different than the everyday definition of selfishness. Genes are built in such a way that they promote their own survival. In most cases this likely means promoting the survival of the organism they belong to, but not always, and the result is not always an organism that behaves in a way that people would normally call selfish. Under the right conditions, an organism's becoming faster, or developing a longer neck, or becoming more active sexually, or even practicing altruistic behavior may be the product of gene selfishness. Vampire bats that eat too much blood during a good hunt will share it with other bats that are hungry. If the gene that causes this behavior spreads widely enough, it increases the same bat's chances of getting food another day, when it has not been as lucky at hunting.

Richard Dawkins lecturing on a book tour for *The God Delusion* (*Mathias Asgiersson*)

Dawkins considers human behavior and even society an extended phenotype—the title of a later book—that works for the survival of genes. He revived Ronald Fisher's idea that cultural elements spread through society in much the way genes do. Dawkins uses the term *meme* to describe these elements, and believes that they are subject to natural selection. Their spread through society can be studied using the principles by which genes move through populations. Logos, sound-bites, blogs, and trademarks are all cultural elements that have been analyzed in this way.

If everything is a part of the extended phenotype, it can contribute to natural selection, so Dawkins disagreed with Gould on the subject of spandrels—he did not think they really exist. Gould responded by saying that genes do not account for everything in evolution. They do not explain, for example, why there are tens of thousands of species of ants, but only a few types of great primates. Second, he pointed out that the genome carries plenty of extra baggage: long repeats of information that seem meaningless, and genes that have lost their functions through mutations. New mutations could make some of this material into new genes, so it is also raw material that can become important for natural selection.

Dawkins has never shied away from making dramatic and very provocative statements in promoting evolution to the general public. Not only is he a self-declared atheist, he sees no real way for religion and science to find common ground. He stated this position clearly in *The Selfish Gene:* "We no longer have to resort to superstition when faced with the deep problems: Is there a meaning to life? What are we for? What is man? After posing the last of these questions, the eminent zoologist G. G. Simpson put it thus: 'The point I want to make now is that all attempts to answer that question before 1859 are worthless and that we will be better off if we ignore them completely.'" Gould and others have taken a less rigid point of view: Certainly genes make human behavior, thoughts about religion, and other parts of culture possible, but this does not imply that the primary function of human existence or society is to ensure that genes can reproduce themselves.

THE EVOLUTION OF HUMAN BEHAVIOR

If nearly everything about an organism is the result of natural selection, then perhaps human behavior and the mind are adaptations to the environment too. Few people would disagree that the behavior of ants, which have tiny brains, is almost entirely directed by their genes. Is the same thing true of human engineering, language, music, mathematics, and science, or are these types of behavior something more?

The aim of the field of *evolutionary psychology* is to understand how the human brain works in terms of its evolution through natural selection. Leda Cosmides and John Tooby, codirectors of the Center for Evolutionary Psychology at the University of California, Santa Barbara, have written a "primer" for this field at their Web site (www.psych.ucsb.edu/research/cep/primer. html). It provides a good definition of the basic philosophy of evolutionary psychology: "The mind is a set of information-processing machines that were designed by natural selection to solve adaptive problems faced by our hunter-gatherer ancestors," they explain. Drawing on information from biology, anthropology, neurobiology, sociology, and other fields, Cosmides and Tooby hope to achieve the same sorts of breakthroughs in understanding the mind that other scientists have made while studying the behavior of birds or other animals, looking for the genetic footprints of natural selection.

Many evolutionary psychologists claim that universal human behaviors such as the ability to use verbal language must have evolved from specific challenges humans faced in past environments. Most of human evolution took place during the hunter-gatherer phase of culture, a period which comprised over 99 percent of the history of humanity. Although people today live in a vastly different world of computers, modern transportation, and urban life, they have to cope using the mental abilities and structures developed during prehistory. Problem-solving strategies have served humans the way camouflage helped other species. This does not mean that everything is determined by genes. Evolutionary psychologists think of the brain as a collection of single modules, something like different cards installed in computers to handle sound, graphics, and communication with other computers. Each module of the brain evolved through different pressures, rather than as an undifferentiated whole, and today they motivate people, help them interpret the world, and are responsible for types of thinking that are found among all humans.

Few disagree with this; the question is what effect it has on daily life. Some psychologists take an extreme view that most human thoughts and actions are somehow encoded in genes. Several of the studies carried out by Cosmides and Tooby suggest that humans have an innate ability to detect deception. Love has an evolutionary value: It helps keep families together for the long periods of time that human children are dependent on their parents. Tastes for particular types of food might have motivated our early ancestors on the African savannah to seek vital elements of their diet that they otherwise would have ignored. Early hominids needed high-energy foods like sugar, salt, and fats, which all were probably difficult to get. Natural selection probably favored people with a craving for them, because they were willing to make an extra effort to obtain these necessary nutrients (for example, by risking being stung when gathering honey). Those taste preferences have been preserved, but in today's society they may lead to excesses that cause obesity, diabetes, and other health problems.

Obviously the building plan of the human brain, its size, tissues, and the way its cells and molecules operate are the result of genes. If the machine does not work right, thinking and behavior cannot happen in the right way. Mutations in specific genes have been found to affect language skills and reasoning, as well as to play a role in learning disabilities, autism, and schizophrenia. Yet few of these cases can be pinned down to single genes, and environmental factors almost always play an important role in whether they develop in a specific person.

On the other hand, studies by Jane Goodall and others have revealed social behavior in primates that seems very human. This suggests that much of human behavior has a long history of selection that stretches far back in evolutionary time. Yet extreme forms of evolutionary psychology have many critics. Stephen Jay Gould criticized the field because most of its conclusions are based on connecting hypothetical causes and effects in ways that cannot be tested. Stephen Rose, of the Free University of London, regards the field as simplistic because it downplays the role that culture, history, and other factors play in people's lives.

The extent to which evolution can explain human psychology is still an open question at the beginning of the 21st century. But as this chapter has shown, 20th-century research made great strides toward resolving many of the major issues with which Darwin and Wallace were unable to deal, such as the mechanisms of heredity and the sources of variation. If evolution had not happened in the way the two scientists believed, genetics and molecular biology would have turned up massive amounts of contradictory data. That was not the case; everything that has been discovered in these fields supports the theory. During the last century, these disciplines were woven together into a version of evolution that explains all levels of life—from single molecules within cells to the shifting genes of huge populations. The next chapters show the role of the theory in today's science.

5

Evolution in the Age of DNA

In the 1950s and 1960s, a revolution in science completely changed the focus of most of biology. Scientists knew that cells were made of DNA, proteins, and other molecules, but they did not understand the functions of these molecules or how they worked together to produce organisms. The discovery that genes were made of DNA completely changed the situation. DNA was the language of evolution, and as scientists learned to decode gene sequences, they suddenly had access to a vast library in which the major steps of evolution had been recorded.

François Jacob (1920–), one of the pioneers of modern biology, said, "When I started in biology in the 1950s, the idea was that the molecules from one organism were very different from the molecules from another organism. For instance, cows had cow molecules and goats had goat molecules and snakes had snake molecules, and it was because they were made of cow molecules that a cow was a cow." Instead, DNA studies have revealed that organisms are built of very similar molecules that carry out similar processes. The tree of life that had been drawn from comparisons of fossil teeth and bones could now be updated. If human genes had more closely resembled those of fish than apes, evolutionary theory would have been dealt a fatal blow. Instead, scientists now had a completely independent method to compare and contrast species. They were also in a position to peer farther back into evolutionary history. In the second half of the 20th century, the new science of molecular biology began to show scientists how to read and manipulate genetic information. This chapter describes how that revolution has influenced the field of evolution.

DNA AND THE MOLECULAR BASIS OF EVOLUTION

In 1953, working in Cambridge in Great Britain, the young American James Watson (1928–) and his British colleague Francis Crick (1916–2004) put together the first chemical model of DNA. They immediately saw the relationship of the genetic code to heredity and evolution.

The two men showed that the four chemical building blocks of DNA, called *nucleotide bases,* are linked into long strands by sugar molecules. The bases of two strands form *base pairs,* making the strands intertwine in the form of a double helix, like the steps and handrails of a spiral staircase. The rail-like strands are the sugars, and the steps are made of a nucleotide in one strand snapped onto a partner in the other. The significant achievement of Watson and Crick was to understand that there are only two kinds of steps. The four bases (adenosine, thymine, guanine, and cystine—abbreviated A, T, G, and C) only form two pairs: an A always binds to a T, and a G to a C. This means that if one strand of DNA has the code AAAA, the other will be composed of TTTT, and the sequence ATATGGC has the complementary sequence TATACCG. It also means that if the two strands are split apart, each contains enough information to reconstruct the second strand. This is what happens when DNA is copied, every time a cell divides. Trying to solve the structure of DNA had occupied chemists and biologists for years. The structure of Watson and Crick was immediately recognized as the correct solution, and also

the answer to the mystery of the composition of genes.

Crick and Watson with their DNA model (*A. Barrington Brown/Science Photo Library*)

DNA is not only a storehouse of hereditary information but it is also an instruction manual that is consulted constantly by cells as they go about their business. It holds the blueprint for making *RNA* molecules, which contain the information needed to make *proteins*. A great deal of biological research has focused on proteins because they control essential aspects of cells such as shape, their attachment to surrounding tissue, and how cells receive and respond to signals from the environment. By carrying out these jobs, proteins transform the hereditary information in genes into processes that build an embryo's body and make it resemble its parents.

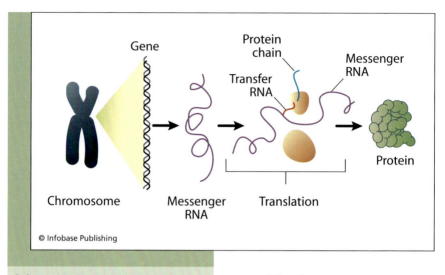

Chromosome   Messenger RNA   Translation

Protein chain   Messenger RNA   Transfer RNA   Gene   Protein

© Infobase Publishing

Information in genes is transcribed into a similar molecule called RNA, which is processed into messenger RNA and then translated into protein by the ribosome.

The human genome carries a huge amount of information; the length of the DNA in each human cell is over six feet (2 m). It can be packed into a tiny cell nucleus because it is so thin. When compressed into tight bundles it takes on the shapes of chromosomes. Humans have 23 pairs of chromosomes, inheriting one of each pair from the mother and the other from the father. With the exception of the sex chromosomes, the two chromosomes in a pair have the same size and shape and usually contain a copy of the same genes. One copy of a gene is usually dominant and the other recessive, which explains why characteristics from each parent are mixed in a child. Except for identical twins, each human being has a unique mix of genes from each parent. This is the main reason for variety in a species.

But new combinations of genes are not the only reason; DNA is never perfectly copied. One kind of change in DNA involves single "spelling mistakes"—known as mutations—which happen as genes are passed along from generation to generation. Other types of change involve rearrangements of parts of the genetic code, in which sequences move to new locations, or the accidental production of extra copies of genes. All of these events are crucial to evolution, and they are discussed later in the chapter.

The fact that mistakes had appeared and were passed along through heredity offered an entirely new way to prove that evolution had happened. Changes could be tracked from ancestral species to their descendants. As scientists decoded the gene sequences of organisms such as the small marine worm *C. elegans,* the fruit fly, and mammals, they discovered that DNA tells virtually the same story as the other methods that have been used to trace the relationships between species. Human genes are more similar to those of chimpanzees than of cows, and more similar to the genes of cows than of bacteria. If evolution had not happened, there is no reason that this should be the case. Jacob's idea that "cows had cow molecules and goats had goat molecules" might have been right. Instead, nearly all cow molecules are also found in goats.

The new science of molecular biology not only showed what genes and mutations are but also promised to show what they do. Darwin had been right after all: Evolution usually happens through incremental changes.

NEW THEORIES OF THE ORIGINS OF LIFE

So little was known about the world of bacteria and other microbes in the mid-19th century that Darwin never ventured detailed guesses about the origins of life on Earth, except to say that it could have arisen from nonliving matter. Chemists of the mid-19th century had started to show that this was theoretically possible: Organic molecules obeyed the same laws of chemistry and physics as nonliving ones. For example, in 1828 the German chemist Friedrich Wöhler (1800–82) synthesized an organic substance (urea) from an inorganic one for the first time. This was the first clear contradiction of a philosophy called *vitalism,* which held that an extra energy or force had to be added to nonliving substances to produce living ones.

In the 1920s, J. B. S. Haldane and Aleksandr Oparin (1894–1980) proposed that life might have arisen through lightning strikes or radiation that changed elements in the environment of the early Earth, which they described as a "primordial soup." Such events could create some of the chemical building blocks of life, which then might self-organize and reproduce themselves. Any complex molecule would undergo errors as it reproduced, making new errors that would open the door for natural selection.

In the early 1950s Stanley Miller (1930–2007) decided to test the Haldane-Oparin hypothesis by trying to simulate the environment of the early Earth in the laboratory. Miller was a graduate student in the laboratory of Nobel Prize winner Harold Urey (1893–1981) at the University of Chicago. Miller assembled a mixture of methane, hydrogen, ammonia, and water in an apparatus consisting of glass flasks and a heating and cooling system. Then he zapped the mixture with electric discharges from electrodes. Urey suspected it might take years for the system to produce interesting results, but within a week, many types of biological molecules had arisen, including several amino acids (the building blocks of proteins). Eventually the experiments produced 13 of the 20 amino acids this way. Over huge stretches of time, the scientists believed that a far richer variety of molecules would be produced, and eventually something that could reproduce itself. The early Earth had at least hundreds of millions of years in which to experiment, and a laboratory the size of the entire planet. (The section "Misinterpretations of Evolution" in chapter 3 calculates the size of this "laboratory.")

Miller and Urey did not know for certain what the early Earth was like, and as ideas about its environment have changed, their experiment has been repeated many times. Amino acids and other "prebiological" molecules arise under widely varying conditions. Such compounds have also been observed in meteorites, the tails of comets, and interstellar gas, leading some scientists (including Francis Crick) to suggest that meteorites and other extraterrestrial material may have seeded the early Earth with some of the basic elements of life. Yet most believe that the Earth was perfectly capable of producing these substances on its own.

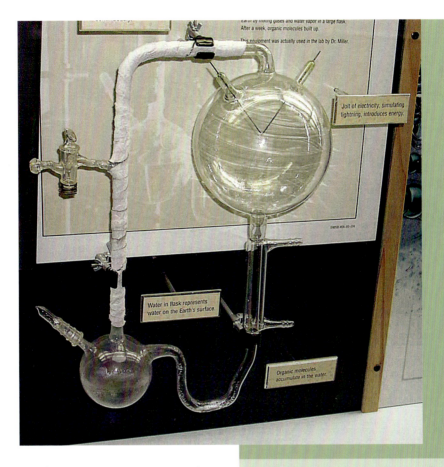

Earth by mixing gases and water vapor in a large flask.
After a week, organic molecules built up.

This equipment was actually used in the lab by Dr. Miller.

Jolt of electricity, simulating lightning, introduces energy.

Water in flask represents water on the Earth's surface.

Organic molecules accumulate in the water.

Stanley Miller's prebiotic soup experiment (*Denver Museum of Nature and Science*)

Scientists are not certain that the first living organism arose in the ocean. Recently some other hypotheses have been proposed. In the 1982 book *Genetic Takeover and the Mineral Origins of Life,* Graham Cairns-Smith presented another scenario: The raw materials of life might have met up and assembled themselves in clay. He pointed out that clay forms crystals whose structure expands by replicating itself, and that sediments at the bottom of streams would be sticky and might serve as meeting points for elements that needed to be assembled to create organic molecules.

A key step on the way to life—and to the beginning of the evolutionary process—was the development of a molecule or a set of processes that could self-replicate. Although amino acids are essential for all life on the planet, they are far from a real living organism and could not have replicated by themselves. It is also unlikely that the first molecule able to self-replicate was DNA. In cells this process requires the help of other very sophisticated molecules that unzip the double helix and read its information. A more likely hypothesis is that the road to life started with something like an RNA molecule, which is closely related to DNA. Because RNA is also built of nucleotides that form pairs, it can also serve as a pattern for making copies of itself. And some RNAs work as enzymes—they carry out chemical transformations of other molecules.

In his 2005 book *Genesis: The Scientific Quest for Life's Origin,* astrobiologist Robert Hazen sums up three current leading hypotheses about the origins of life. He calls the RNA hypothesis a "genetic scenario" in which sophisticated, information-loaded molecules arise and grab the elements they need to copy themselves from the environment. "Metabolic" hypotheses propose that life began as a sequence of chemical processes that assembled themselves on a surface, like a mineral, and that features like self-replication and a primitive membrane were added through a process of natural selection. A strong point of this idea, Hazen says, is that it begins from the point of view that life needs energy, which could be provided by minerals. "A third scenario," he writes, "rests on the possibility that neither protometabolic cycles (which lack the means of faithful self-replication) nor protogenetic molecules (which are not very stable and lack a reliable source of chemical energy) could have progressed far by themselves. If, however, a crudely self-replicating genetic molecule became attached to a crudely functioning surface-bound metabolic coating, then a kind of cooperative chemistry might have kicked in." This "dual origins" hypothesis is more complex, Hazen admits, and may not have been the first step in life's origins, but in any case, the two types of processes probably happened very early in evolution.

Evolutionary theory itself does not deal with the origins of life—the transformation of inorganic substances into organic ones—but today's chemists point out that the principles of variation and natural selection apply to any chemical system that replicates itself imperfectly. The first self-replicator was not alive, but it used natural materials to reproduce and, in the process, changed its environment. It linked up to other molecules and then faced the problem of keeping the pieces together. One hypothesis is that everything may have assembled near thermal vents in the ocean, gathering in honeycomblike pores in rocks around the vents, or in other natural structures that did something similar. The solution that cells use today, wrapping themselves up in membranes, was a later development.

Membranes are made of highly complex fat molecules called *lipids.* Scientists do not yet understand how they arose from inorganic substances. At some point they became protective envelopes that could shield other molecules from aggressive chemical activity in the environment that might break them down. Membranes also acted as surfaces that molecules could be attached to, allowing them to be organized in space. This could increase their efficiency and reproduction, just as there are optimal positions for the machines in a factory.

Although the idea that life began as RNA molecules has gained in popularity, scientists do not yet know how double-stranded DNA could have arisen from single-stranded RNA. Recently a French researcher named Patrick Forterre proposed a hypothesis that viruses might have been involved. In a 2005 paper in the journal *Biochemie,* Forterre wrote, "It is very likely that cells and virus-like organisms already coexisted and fought each other (or cohabited in various ways) in the RNA world." He points out that many researchers believe that genomes made of RNA existed before cells with DNA. The Earth currently supports an unbelievable number of viruses—these tiny collections of genes, RNA, and other molecules make up most of the biomass of the ocean. A similar situation probably existed during the first phase of life on Earth. Viruses sometimes kidnap genetic material from one cell and transfer it to the next. Once DNA existed, Forterre believes, viruses might have passed it around. It is more stable than RNA and could have won out over RNA-based life through competition and natural selection.

The first living organism was likely the product of hundreds of millions of years of chemical evolution; still, it seems to have evolved relatively early in the Earth's history. The oldest rocks known on Earth, found in Greenland, are about 3.8 billion years old; some chemists believe they may have already contained living organisms. Clear traces of fossil cells have been found at various sites dating from 3 to 3.5 billion years ago. By that time, life was probably widespread.

It is impossible to tell whether these fossils represent the direct ancestors of humans or other forms of life on Earth today. In the early days of our planet, many types of biological systems may have arisen. Only one of them seems to have survived: an organism known as the *Last Universal Common Ancestor* (LUCA) of everything currently alive on Earth. No direct traces of it have been found, but this cell has left its mark on every living creature. Scientists are beginning to be able to reconstruct some of its features, based on analyses of genomes that are discussed in chapter 6.

The descendants of LUCA took three different evolutionary paths to become the great branches of life on Earth today: bacteria, *archaea,* and *eukaryotes.* The last group includes all plants and animals as well as a wide range of one-celled organisms such as yeast. The main difference between eukaryotes and the other branches is their extra compartment, the nucleus, which contains most of the cell's genetic material and acts like a cell within a cell—it may even have started off this way, as one organism that infected another. The membrane around the nucleus offers extra protection for DNA, and that shield was probably essential to the creation of huge genomes.

Archaea were first discovered a few decades ago in hot springs and thermal vents in the ocean, and were thought to be bacteria until studies of their genes and biology revealed significant differences. In the meantime they have been found all over the Earth. Some species still live at very high temperatures or in other extremes, such as in environments with high concentrations of salt that would be deadly to other forms of life. Studies of archaeal genes reveal that they are probably the oldest existing form of life, making bacteria and eukaryotes their descendants.

A number of recent studies provide evidence that the earliest bacteria to evolve also likely lived in hot environments. In a

2008 article in the journal *Nature,* Omjoy Ganesh of the University of Florida showed not only that several important families of ancient proteins worked well in a hot environment, but that their structures adapted progressively as the Earth cooled. Between 3,500 and 500 million years ago the Earth's temperature dropped by about 54°F (30°C), a change which is recorded in the genomes of organisms alive today. So LUCA was probably a *thermophile.*

Now the Earth's most widespread form of life, bacteria can live almost anywhere, from the insides of our bodies, to oil wells, to the polar icecaps. They are so hard to get rid of that scientists believe they might survive long space voyages, even exposed on the surface of a ship, so satellites and probes undergo extensive sterilization procedures to avoid sending Earth life to other planets.

FROM SINGLE CELLS TO MULTICELLULAR ORGANISMS

Until recently, archaea and bacteria were thought of as single-celled organisms that occasionally lived together in loose colonies. It was believed that true multicellular life only evolved in the eukaryotes. Recently new forms of archaea have been found that seem to live together in long-lasting groups in which single cells take on specialized functions. Researchers are also coming to a new appreciation of the group behavior of bacteria, which form moving sheets that can invade new environments. In hospitals, for example, they can crawl up catheter tubes and cause infections. Thus even bacteria change their behavior to take on specialized functions in a community.

The formation of colonies has helped single cells survive since the very early history of life. Fossils dating back over 3 billion years seem to indicate the existence of large, matlike collections of cells. Similar colonies exist today, although they were much more common in the ancient past. Scientists do not consider them true multicellular organisms, which are usually defined as creatures that undergo several phases of development and whose cells become dependent on each other by differentiating into distinct types.

The eukaryotes are divided into the kingdoms of plants, animals, and three more: fungi, red seaweeds, and brown seaweeds. All of these organisms stem from a single ancestor whose evolutionary history stretches back much further and is itself very interesting. In the 1970s Lynn Margulis (1938–) of Boston University proposed a hypothesis to explain why cells have an unusual compartment called the mitochondria. This structure possesses a large collection of its own DNA that reproduces independently from the genetic material in the cell nucleus. Margulis claimed that it arose when an early bacteria invaded a cell (or was eaten by it) and stayed, forming a successful partnership with the larger cell. The two organisms eventually became completely dependent on each other.

The Russian scientist Konstantin Mereschkowsky (1855–1921) had proposed something similar at the beginning of the 20th century but did not have the methods to build a convincing case. Recent comparisons of the DNA of bacteria and mitochondria, as well as other types of evidence, support the idea. If they did begin as separate organisms, mitochondria have adapted fully to symbiotic life, serving the cell by breaking down carbohydrates to produce energy.

Recently a more controversial hypothesis has been put forward proposing that the nucleus of eukaryotic cells began as an archaea. And structures in plants called *plastids* (such as chloroplasts, which are responsible for photosynthesis) may also have arisen through a symbiosis of separate cells.

There are many one-celled eukaryotes, and their genomes are amazingly diverse. The one-celled amoeba called *Paramecia* and the *Plasmodium* parasite (which causes malaria) are genetically more different from each other than a human being and a daffodil. At some point—probably about 1.5 billion years ago—a

(opposite page) One-celled marine creatures called choanoflagellates (above) are nearly identical to a type of cell called a choanocyte, found in sponges. Studies of these organisms are providing insights into the rise of multicellular life.

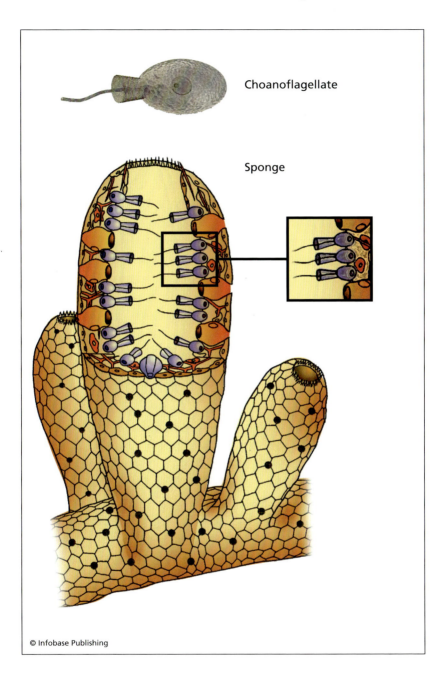

Choanoflagellate

Sponge

© Infobase Publishing

small group of eukaryotes evolved into true multicellular life. Some very odd creatures alive today give some intriguing hints about how this might have happened. A slime mold (a type of fungus) called *Dictyostelium* lives as a single-celled amoeba until its food supply gets scarce. Then several hundred of the cells draw together to form a wormlike creature, which crawls along until it finds a richer supply of food. Once there, it begins to reproduce. First the worm transforms itself into a long stalk topped by a bulbous head, then the head bursts, scattering new single-celled organisms into the environment. The first animals and plants may have behaved in a similar way, spending part of their time as individual cells, then joining into groups and taking on specialized forms and functions for a part of their life cycle.

Other eukaryotes also seem to straddle the line between unicellular and multicellular life. The green algae volvox forms hollow balls that spin through the water. These come in various sizes, from several hundred to tens of thousands of cells that undergo specialization. Most of the cells make up the surface, but a small number are set aside to reproduce.

Sponges also have fascinating characteristics that say something about the origins of animal life. Their digestive tissue is lined by cells called choanocytes, each of which consists of a long hair-like cilium surrounded by a sort of fringed collar. The cilium beats and draws water in through the collar. As that happens, small particles get caught in the fringe; these provide the sponge's food.

Choanocyte cells are particularly interesting because they are closely related to single-celled organisms called choanoflagellates, which look nearly identical and behave the same way. (They are also extremely similar to the cells that make up volvox.) Choanoflagellates are thought to be the closest living relative to the ancestor of all animals, and adult sponges seem to be little more than a skeleton to support choanocytes. They are an amazingly successful form of life: Fossil sponges have been found that date back more than 600 million years, and over 5,000 species exist today.

READING HUMAN EVOLUTION FROM FOSSILS

Until very recently a few sparsely scattered bones were the only way to understand most of human evolution. The situa-

Major Eras in the Development of Plant and Animal Life

Scientists traditionally divide the history of life on Earth into 17 or 18 major periods, based on geological and fossil records. Periods are usually linked to the appearance of different forms of life and often end in mass extinctions. Some of the major eras are listed below, along with key evolutionary events that took place.

- The Precambrian (from the origins of the Earth to about 540 M.Y.A.): The first living organisms appeared, likely between 4 and 2.5 billion years ago, probably developing into archaea, bacteria, and eukaryotes by 1.6 billion years ago. The eukaryotes developed into multicellular life forms such as algae, sponges, jellyfish, and worms. The end of this period saw the origins of the first "bilateria"—animals with symmetrical body plans, likely somewhat similar to a small worm called *Platynereis dumerilii,* which is alive today.
- The Cambrian (540–500 M.Y.A.): This period saw an "explosion" of new types of multicellular species with a huge variety of new shapes and structures. Nearly all of the basic types of body plans of today's animals developed within this period. These include mollusks (which would lead to clams, snails, slugs, squid, and octopuses), annelids (which would become earthworms and leeches), arthropods (insects, spiders, centipedes, and crustaceans), brachiopods (marine invertebrates with hinged shells), and echinoderms (like sea urchins and sand dollars).

(continues)

(continued)

- The Ordovician period (500–435 M.Y.A.): The first plants developed on land, and arthropods (insects) climbed out of the sea.
- The Silurian period (435–410 M.Y.A.): Mosses, fungi, and jawless fish developed.
- The Devonian (410–355 M.Y.A.): The evolution of land-based predators (huge, scorpionlike creatures), jawed fish, and cephalopods (like the octopus). The first four-legged vertebrates (tetrapods) moved from the sea to land, evolving out of a type of fish with lobed fins.
- The Carboniferous (355–295 M.Y.A.): The formation of most of the Earth's coal deposits, as the sea rose and receded, covering forests of ferns, mangroves, and many types of trees. The fossils of dragonflies with wingspans of about three feet (1 m) have been found in these geological layers, as well as monstrous millipedes. Up until this time, land-living tetrapods had their offspring in the water, but now creatures developed eggs with shells and a yolk, or carried their embryos in an amniotic sac.
- The Permian (295–250 M.Y.A.): Forests of seed plants developed, as well as a flying lizard. This era ended with an enormous mass extinction of plant and animal species.
- During the next three periods, lasting nearly 200 million years, dinosaurs ruled the land and the

tion has changed considerably now that molecular biologists have completed the genomes of humans and their closest relative, the chimpanzee, and scientists are currently completing the sequences of many more animals. Studies of humans are a seas. The first dinosaurs appeared in the Triassic (250–200 M.Y.A.), and this era also saw the development of mammals, frogs, turtles, and crocodiles. Many of today's common creatures evolved during the Jurassic (200–135 M.Y.A.): birds, lobsters, sharks, and major families of insects (flies, bees, ants, butterflies, and moths). Stegosaurs and allosaurs appeared. During the Cretaceous (135–65 M.Y.A.), snakes, lizards, termites, and loons made an appearance alongside enormous dinosaurs like *Tyrannosaurus rex*, the Triceratops, and flying pterasaurs. The end of this period is marked by massive extinctions, accompanied by evidence of a huge asteroid striking the Earth, and periods of cold and massive fires which swept the planet. Survivors included plants with seeds, birds, and many of the mammals.

- Several eras spanning from 65 M.Y.A. to about 1.75 million years ago saw the development of grasslands, savannahs, and modern forests. At the end of the Pliocene (5.3–1.75 M.Y.A.), hominids diverged from chimpanzees and gorillas.

Scientists have been assembling this history of life since the discovery that geological strata reflected successive eras in the history of the Earth and that they held unique sets of fossils. Exact dating has been possible thanks to the development of methods that measure the progress of radioactive decay in minerals.

good example of the limitations of past methods of studying evolution and the promise of new ones.

Darwin hypothesized that humans were quite closely related to chimpanzees and that these species likely originated in

Africa. Both fossil evidence and modern genetic data support these claims. As scientists studied the first data from genes from humans and chimps, most were surprised to discover that the two species had diverged very recently—likely between 5 million and 7 million years ago. This seemed like a short time for humans to have developed such unique characteristics.

Classifying humans has always posed a problem for biology. Walking upright, making tools, using language, and unique features of the brain emphasize the differences between humans and other animals. These overshadow the similarities which are being revealed by genome studies: Humans and chimpanzees are far more alike than some species of sponges. While this fact will not make much practical difference in people's daily lives—it will not change the way they react to art or music, or how they play football—it has already led to new insights into human biology and genetic diseases.

Darwin's principle of common descent states that the similarities between two related species are almost always inherited from their common ancestor. If that ancestor no longer exists, some of its features can be reconstructed by comparing the traits shared by its descendants. A great deal has been learned by comparing human and chimpanzee DNA. It is not yet possible to say exactly what the common ancestor of these species looked like—or to answer questions such as when *hominids* were first capable of language—but progress on those questions may come soon, as scientists learn more about the connection between genes and the building plans of the brain and the rest of the body.

In line with Darwin's hypothesis, remains of the earliest known hominids have been found in Africa. Geology and fossils show that the continent underwent significant climate changes several million years ago, as dense forests replaced open woodlands. Later these became grassy, open savannahs. New types of primates evolved that were suited to life on the open plains. They began to walk upright, and later their brains became significantly enlarged.

Until very recently, the oldest known hominid of this type was *Ardipithecus ramidus,* found in 1994 by University of Califor-

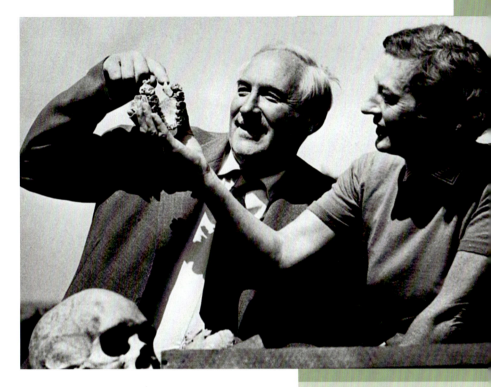

nia, Berkeley, researcher Tim White on top of a layer of volcanic ash in Ethiopia. The ash permitted the specimen to be dated at roughly 4.4 M.Y.A. A milk tooth and other features of its teeth set it apart from apes and a later type of hominid called *Australopithecus*. Other features, such as fragments of the arm and the base of the skull, also have "in-between" characteristics. It might have

Louis and Mary Leakey. Darwin's hypotheses about the origins of humans and early hominids spurred the Leakeys and other paleontologists to mount expeditions to Africa, where they found a wealth of fossils documenting early hominid history. *(The Leakey Foundation)*

lived before the split of the human lineage and chimpanzees, and even been a close relative of their common ancestor.

Then in 2001, French scientists discovered a skull in Chad, Africa, which they believed to have belonged to a much older hominid. In 2008 Michel Brunet, who had christened the species *Sahelanthropus tchadensis,* used radiometric dating to show that the strata in which the fossil had been found was between 6.8 and 7.2 million years old. Three-dimensional computer reconstructions of the skull revealed features characteristic of upright-walking early hominids. Brunet and his colleagues concluded that the fossil may well represent a very early ancestor of both chimpanzees and humans. In the meantime, this interpretation has been widely debated by scientists. In 2006 Milford Wolpoff, a paleoanthropologist at the University of Michigan, pulled together the evidence in an article "An Ape or *the* Ape: Is the Toumaï Cranium TM 266 a Hominid?" published in the journal *PaleoAnthropology*. Wolpoff believed that Brunet had used an outdated method to interpret the teeth, and also that the structure of the base of the skull (which showed that *Sahelanthropus* walked upright) had been misunderstood.

Wolpoff concluded "*Sahelanthropus* was an ape living in an environment later abandoned by apes but subsequently inhabited by australopithecine species. . . . Yet, it is a highly significant discovery . . . perhaps mostly because of the insight it might give for understanding the ancestral condition before the hominid-chimpanzee split." Even if it is not the common ancestor of humans and chimpanzees, he writes, "It does provide important information that helps us better understand this divergence." It is also an excellent example of how scientists go about reconciling the information obtained from fossils and molecular data.

These finds follow a long line of important discoveries in Africa. Seven decades earlier, Raymond Dart (1893–1988) had discovered remains of a hominid skull with humanlike features in Cape Province; he named the specimen *Australopithecus africanus* (for "ape of southern Africa"). The fragments were likely between 2 million and 3 million years old. In 1974, a team of paleontologists in Ethiopia headed by Donald Johanson (1943–) discovered a more complete set of similar remains, which they named *Australopithecus afarensis*. Enough of the skeleton's pelvis was found to prove that the hominid must have walked upright. Shortly after that, Mary Leakey (1913–96) and her colleagues found a set of 3.6-million-year-old footprints from an upright-walking creature in Tanzania. It might have been *A. afarensis*.

These two species and a third, called *Australopithecus garhi*, belonged to a lineage that is known as the gracile line of hominids because of their light bones and jaw structure, compared to robust counterparts with a much thicker body structure and heavier jaws. The gracile hominids probably evolved into *Homo habilis,* which is recognized by nearly all experts as the ancestor of modern humans. *A. garhi* remains have been found in sites with stone tools. The same is true for *H. habilis,* who lived about a million years later. Examinations of *H. habilis* skulls show that their brain, while about half the size of the modern human brain, may have had a structure like the Broca's area, which contributes to language abilities in modern humans.

At some point, hominids migrated out of Africa into Asia and the Indonesian islands. Fossils of *Homo erectus,* a successor to *H. habilis,* have been found in sites as far away as China, and archaeologists have found stone tools on a Pacific island called Flores that seem to have been made by *H. erectus* over 800,000 years ago. If so, *H. erectus* must have been able to build boats, because Flores could only be reached by water. The hominid had other impressive skills: Recently a finely-balanced spear was found in Germany that was made a million years ago.

Most scientists believe that *H. erectus* was a cousin, rather than a direct ancestor, of modern humans. This species was probably similar to *Homo heidelbergensis,* also known as "archaic homo sapiens," whose remains have been found in Europe, Africa, and parts of Asia. *H. heidelbergensis* may be the ancestor of both humans and another cousin, Neanderthal man. These species, too, migrated from Africa to settle large parts of the world.

More than 100,000 years ago, a new species called *Homo sapiens* emerged, probably once again in Africa. These were the first modern humans, and they were equipped with brains that had developed to become twice the size of those of their early australopithecine ancestors. They cultivated plants and created art and fabulous tools. All that is known of the fate of other hominid species is that they became extinct.

For several decades rsearchers placed the origin of modern humans at about the 100,000-year mark, but in 1997 a team headed by Tim White made another important discovery in Ethiopia. They found the skulls of two adults and one child that could be dated back to about 160,000 years. Not only does the discovery push back the origins of *Homo sapiens* by tens of thousands of years, but it also helps confirm the hypothesis that the species originated in Africa.

FOSSIL DNA

By the mid-1980s, scientists were actively reading the sequences of genes, and they wondered whether DNA could be extracted from old specimens, perhaps even fossils. At the end of the decade, Svante Pääbo (1955–   ) at the University of Münich proved that traces of DNA could be found in well-preserved ancient tissues, even mummies that were thousands of years old.

This was possible thanks to a revolution in biotechnology in the late 1980s that made it possible to easily "amplify" DNA— basically putting samples of genetic material into a chemical "copy machine" and producing billions of identical molecules. The method, called the *polymerase chain reaction,* or PCR, had a huge impact on medicine, molecular biology, and the field of ancient DNA, because it gave scientists a way to amplify the DNA even from the tiny amounts of samples that could be obtained from fossils.

The 1990s saw many laboratories attempt to do this. Journals published sequences that were supposedly obtained from fossil plants, organisms preserved in amber (especially insects), and even dinosaur eggs. This inspired books and films such as *Jurassic Park* and suggested that it might be possible to trace evolution by directly reading the genetic code of extremely old fossils. Unfortunately, most of these incredible early findings turned out to be false, the result of contamination in the field or the laboratory. (One dinosaur DNA sequence turned out to be that of a lab technician.)

In the meantime, methods have been vastly improved. Pääbo and his colleagues now have learned to handle and amplify fossil DNA under extremely sterile conditions. They have also discovered that there is probably an upper limit of about 100,000 years on the age of fossils from which DNA can be recovered. Over time, DNA molecules undergo chemical transformations that destroy them or break them into such tiny fragments that it is impossible to obtain meaningful information from them. Jurassic Park will never become a reality if it depends on recovering DNA from dinosaur remains, or from mosquitoes trapped in amber.

Some of the most interesting methods of tracking human evolution involve analyzing DNA from the Y chromosomes of men and another type of DNA contained in small cellular structures called mitochondria. Females do not have Y chromosomes, so men always inherit them from their fathers and pass them down to their sons. By analyzing small regions of the chromosome from men across the world, biocomputing experts have determined that all men can trace their genetic history back to a single individual who lived in Eastern Africa from 60,000 to 90,000 years ago. This *most recent common ancestor* of men has been nicknamed "Y chromosomal Adam." People also have a type of DNA that they inherit uniquely from their mothers, in the small cellular structures called mitochondria (see "From Single Cells to Multicellular Organisms" above). Most cells have hundreds or thousands of these organelles, which reproduce independently of the cell nucleus, and they all come from the mother's egg cell. Studying variations in their DNA has revealed that humans also have a female most recent common ancestor. Scientists have nicknamed her "Mitochondrial Eve," and she also lived in Eastern Africa, sometime between 140,000 and 200,000 years ago.

The DNA of existing people and organisms is one method of probing the evolutionary past; additionally, scientists are discovering ways to increase the amount of information they can retrieve from fossils using enzymes that repair broken fragments of DNA.

This strategy has been put to use to study the remains of humans and animals who died in the destruction of Pompeii in the year 79 C.E. In one study, a team of scientists from the University of Naples obtained samples from the bones of people who had barricaded themselves in a house to wait out the deadly rain of stones and ash falling on the city. The arrangement of the bodies suggested that some of them, at least, might belong to the same family. The bones were preserved well enough that some fragments of DNA could be extracted and analyzed. The scientists are currently studying the data, hoping to discover something about the familial relationships of the victims. Preliminary results have hinted that one of the victims appears to be of a different ethnic heritage, possibly a slave captured during one of the Romans' many wars in foreign lands.

A study of DNA from horses in a Pompeii stable revealed that the Romans bred types of animals that no longer exist, including unique mixtures of horses and donkeys. Many other studies have now been conducted on ancient DNA obtained from ancient sites around the globe. Examinations of tissues from mummies found in Egypt and South America turned up traces of the tuberculosis bacterium, showing that this fatal disease is probably as old as modern humans, possibly even older.

In 1997, Svante Pääbo announced that his laboratory in Munich had obtained DNA samples from the bones of a famous specimen: the first recognized example of Neanderthal man, discovered in Germany in 1856. The samples provided enough information to make a meaningful comparison to the human and chimpanzee genomes and finally put an end to a century-old debate: Were Neanderthals the ancestors of modern humans?

Neanderthals lived throughout the area that is modern Europe, Asia, and many other parts of the world from about 300,000 to 30,000 years ago. They used tools and carried out ritual burials, and during much of this time they lived alongside modern humans. Then they completely disappeared, and no one knows why. Since their lifestyle and behavior strongly overlapped with that of modern humans, competition for resources may well have played a role in their demise. Another possibility is that the Neanderthals succumbed to diseases carried by modern humans, who likely brought a wide range of unfamiliar parasites with them as they migrated from warmer tropical regions of the globe. History is full of examples of plagues that arise when modern human cultures have been separated for thousands of years and then come into contact. When Spanish explorers arrived in the New World, for example, they brought along smallpox, which wiped out millions of Native Americans. It is likely that the Neanderthal and modern human bodies were similar enough that viruses or other diseases could have been transmitted from one species to the other, with devastating consequences. If diseases passed between the two species, they might have learned to avoid each other, and this might also explain the lack of archaeological evidence for direct contact or cohabitation between them.

While most evidence suggested that Neanderthals were cousins rather than the direct ancestors of modern humans, the exact relationship had not been entirely resolved. A few scientists believed that the two species might have been capable of breeding successfully with each other, meaning that some Neanderthal genes might still survive in modern man. However, Pääbo's analysis of fossil DNA has revealed such marked differences that this is very unlikely; in fact, Neanderthal genes are more closely related to those of the chimpanzee than to those of *Homo sapiens.* In the meantime, Pääbo and his colleagues have obtained DNA from the fossils of several other Neanderthals from different parts of the world. These studies confirm the original findings.

6

Evolution in the Age of Genomes

The 21st century opened with a huge scientific landmark: the completion of the human genome. By the time this book appears, scientists will have obtained complete DNA *sequences* of several hundred organisms, including dozens of other complex animals. This information has provided fascinating new insights into how genomes have changed over billions of years; at the same time, it is answering fundamental questions about human biology by comparing the genes of humans to those of other organisms. This is only possible because of the links between humans and other species. This chapter gives a brief overview of what scientists have learned about evolution in the age of genomes, and what they hope to discover in the near future.

Until the 1990s sequencing DNA was a slow, difficult process, which meant that researchers usually had to be satisfied with analyzing genes one by one. Scientists tracked evolution's footprints by reading the information in single genes. They chose big molecules with hundreds or thousands of nucleotide bases, long enough to preserve a detailed record of the history of mutations. As interesting as this was, it still was a bit like watching the evolution of a single tooth, without considering an entire organism. The most interesting questions about evolution could likely only be resolved by studying entire genomes.

By 1990 the invention of a method called *capillary electrophoresis* and other improvements in biotechnology had greatly sped up the process of obtaining DNA sequences. Researchers began working on genomes, starting with those of small viruses. The first free-living cell was sequenced in 1995—a bacterium called *Haemophilus influenzae,* which causes meningitis. This was followed a year later by the first eukaryote—baker's yeast—and the first complete animal, a worm called *C. elegans,* in 1999. Finally in 2001, the entire human genome sequence had been read by the American company Celera and dozens of laboratories collaborating in the International Human Genome Sequencing Consortium.

Obtaining a complete genome sequence is like having learned the alphabet; reading the information it contains is an entirely new challenge. A genome is simultaneously a recipe book and a toolbox. Scientists would like to know more than how the spellings of genes have changed over time; they hope to learn how these changes alter the bodies and lifestyles of living creatures. But this can only be done with a complete list of genes and very detailed information about how they function together in cells and organisms.

FROM THE EVOLUTION OF GENES TO GENOMES

Comparisons of complete genomes have turned up some surprising facts both about our own species and the history of life. About 21 percent of human genes have close relatives in all other animals, plants, bacteria, and archaea; these molecules existed in the last common universal ancestor of life on Earth, and may be over 3 billion years old. By the time eukaryotes split off from archaea and bacteria a billion years later, 53 percent of human genes already existed. With the first animals, probably about 1.5 billion years ago, 75 percent of those genes were already in place, and the evolution of the first vertebrate brought nearly all of the rest—99 percent, which have existed for about

500 million years. Fewer than one percent of human genes are truly unique.

The human genome has about 3,165,000,000 bases. Genes account for less than 2 percent of that, which is equally a product of evolution. The genome is littered with artifacts called *pseudogenes,* the remains of ancient genes that have lost their functions through mutations. For example, an ancient ancestor to both humans and mice had at least 1,000 genes devoted to smell. Of those, 20 percent have lost their functions in the mouse, and over 60 percent have been lost in *Homo sapiens.* (Apparently such acute powers of smell are no longer very helpful to the survival and reproduction of humans.) Similarly, a molecule called MYH16 is found in the chewing muscles of most primates. Humans have lost the function of this gene—possibly as the structure of the faces and its muscles changed.

Genes are not like muscles; they are not lost because they are not used. A better way to think of the situation is this: Random events are likely to make any gene disappear unless it is needed. Mutations are likely to occur at an equal rate almost anywhere in the genome, and they usually damage or disable genes. Changes in gene sequences often change the chemistry and functions of proteins, which means that cells are unable to create functional molecules. If the protein is not very important, an organism may survive and reproduce equally well without it. But damage to a crucial gene will probably reduce the chances that an animal's genes get passed along to the next generation. So indirectly, natural selection creates a "use it or lose it" situation for genes.

Some DNA sequences are not genes—they do not encode proteins—but influence their behavior. For example, they serve as docking sites for proteins that activate nearby genes. Even these "control elements" represent only a small part of the genome. Nearly half of the human genome consists of small "transposable elements," fragments that copy themselves and then get sewn back into the DNA, like inserting additional links into a chain. An additional quarter of the genome consists of bits of code called *introns* which are found in the middle of genes. They are transcribed into RNA, but are removed before the RNA is used to make a protein.

The different types of DNA sequences give researchers various types of "counters" to measure the rate of evolution. Mutations in genes are not allowed too often, so they happen at a slow pace, like the hour hand of a clock. Extremely important proteins permit almost no mutations, so relationships between genes can be detected at vast evolutionary distances, such as those between bacteria and humans. Noncoding DNA sequences undergo mutations at a much faster rate, like the second hand of a clock, so they are useful for looking at close relationships—between humans and chimpanzees, for example. They can also be used to detect family relationships within a species. And several hundred noncoding regions of DNA scattered throughout the genome contain repeated bits of code that evolve much more quickly. These regions, called *microsatellites,* change so quickly that scientists can use them in *DNA fingerprinting:* a method to match DNA samples to single individuals or to determine family relationships between people.

Comparisons of genomes have also revealed that many regions of noncoding DNA are evolving more slowly than at a random rate, which means that they likely have uses in the cell. New findings are revealing some of those functions. One discovery has to do with the fact that DNA is made up of two strands—known as "Watson" and "Crick." Some genes are located on the Watson strand and others on the Crick strand. Until recently, scientists have usually assumed that the information on the strand directly across from a gene is meaningless. But new studies of cells have revealed that the second strand is widely used to produce *microRNAs.* These small molecules do not encode proteins, but sometimes have other functions in the cell. They may, for example, dock onto another RNA and cause it to be destroyed. Then it cannot be used to synthesize proteins. The next few years are likely to reveal many more functions for these and other noncoding regions of the genome.

GROWING AND SHRINKING GENOMES

Mutations usually involve changes in single letters of the genetic code, but genomes constantly undergo much more dramatic changes, acquiring and losing large blocks of genes, sometimes even entire chromosomes. Some organisms have lost incredible numbers of genes, shrinking down to a tiny size—direct evidence that evolution does not always make organisms more complex. Simplification is especially common among infectious species of bacteria. If they spend their entire lives inside a plant or animal, they can "outsource" many of their needs to the host's cells. In doing so, they sometimes become more dangerous to the host. The bacterium *E. coli* has over 4,000 genes. It usually lives in the human gut without doing much damage. But the simpler organism *Shigella,* its close relative, causes dysentery. Another related bacteria, *Buchnera,* has whittled its genome down to only 590. That number still permits it to survive in the cells of aphids. Plant and animal species have also lost significant amounts of DNA.

Evolution can simplify the structure of genes as well as reduce their number. The last section introduced introns—noncoding DNA sequences within genes. The simplest genes have no introns; complex genes may have hundreds. In 2005 the lab of Detlev Arendt at the European Molecular Biology Laboratory, in Germany, showed that the genome of the fruit fly is not only quite a bit smaller than the common ancestor of insects and vertebrates, but its genes are also less complex. The average human gene contains 8.6 introns; in the fruit fly, the average is only 2.3 per gene. Originally scientists believed this meant that genes have been getting more complex over the course of evolution—humans are more complex than flies, by most ways of measuring things, and thus are often thought of as "more evolved."

Arendt points out that humans are evolving at a much slower pace than insects, partly because of the speed at which insects reproduce. Fruit flies reach sexual maturity two weeks after birth, compared to 14 or 15 years in humans (who usually wait longer than that to reproduce). Thus, in just 250 years flies have given birth to about the same number of generations as humans have produced in 100,000 years. Since every new generation is an opportunity for mutations and other changes in DNA to creep into the genome, flies are evolving faster.

Arendt's study compared the introns of humans, flies, and a worm that resembles their last common ancestor. His conclusion was that ancient genes were more like those of humans than flies—they had more introns. Here evolution has not been making things more complex; it has been simplifying things by cutting out introns in fly genes.

While the genomes of bacteria and flies have been shrinking, those of vertebrates have expanded tremendously. Researchers think this has happened in two ways. Sometimes when DNA is copied, small regions get duplicated extra times—like pushing the button on a photocopy machine twice. Recent studies of genomes have shown that this happens nearly as often as mutations. This may give an organism two copies of the same gene. Initially the copies are identical and often behave the same way in cells, leading to the synthesis of double amounts of a protein. As time goes on, the two copies undergo different random mutations that can have various effects. One molecule usually retains the old jobs. The other may be damaged and disappear, or it may undergo changes that give it new tasks. For example, it might be produced only in specialized tissues, which can give cells new features and eventually produce new organs.

In the late 1960s, Susumu Ohno (1928–2000) made the claim that *gene duplications* have been the most important factor in evolution. The human genome is full of examples, such as nine functional globin genes. Cells produce globin molecules and link them in groups of four to make hemoglobin, the molecule in blood cells that carries oxygen. At different times the body creates different sets of globins to do this, which slightly changes how it uses oxygen. Embryos create unique sets of globins that carry more oxygen. Multiple copies of the genes are also found in birds, and in all of these species they have undergone unique types of specialization. A bird called the Rüppell's griffin builds four different types of hemoglobins, giving it the amazing ability to fly at altitudes of 10 km (six miles) above sea level.

The sophisticated immune system of humans and other mammals can largely be traced to duplications of immunoglobulin genes. The human genome contains over 150 immunoglobulins, which can be linked together in a practically infinite number of

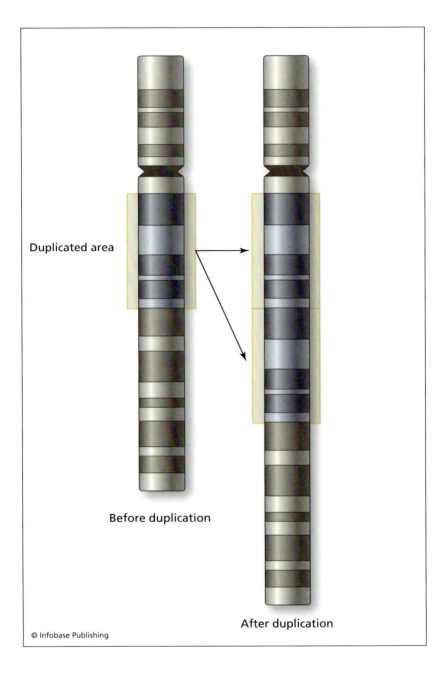

Duplicated area

Before duplication

After duplication

© Infobase Publishing

patterns. These are used to create antibody molecules that have random patterns, which is like cutting billions of randomly shaped keys; eventually one of them will fit a lock. If an antibody hap-

(opposite page) One type of error that occurs during cell division is that extra copies may be made of single bases, small sequences, huge regions of chromosomes, or even entire genomes. This provides more "raw material" that may acquire new functions through evolution.

pens to match a molecule on the surface of a virus, it binds to the virus and alerts the immune system, which disposes of the problem. This type of immunity has allowed humans and other mammals to survive when they move into new environments, which always brings challenges from new viruses and microorganisms.

Gene duplications gave primates a system of vision based on three colors, and they have also greatly enhanced mammals' sense of smell by spinning off new receptor molecules in sensory cells in the nose. The number of such olfactory receptor genes has gone down again as humans have lost many of these genes, compared to chimpanzees. Interestingly, some subtypes of smell genes have been maintained in humans. Yoav Gilad and Svante Pääbo's laboratory (now in Leipzig, Germany) thinks the fact that humans cook their food might explain both the loss of some types and the preservation of others.

Vertebrates often have three or four copies of genes that are found only once in other branches of life. This led Ohno to propose that the entire genome had been duplicated in an ancient organism at least once, and possibly twice. This is not hard to imagine: When cells divide, the entire genome is copied, and then the two sets are equally divided into separate compartments. If something goes wrong with the compartmentalization machinery, one cell could be left with two copies. This can happen with the entire genome or single chromosomes; such "sorting problems" lie at the root of diseases like Down syndrome, where a fertilized egg receives three copies of chromosome 21. Whole genome duplications are known to have happened in plants such as wheat and rice, possibly as a result of humans selecting plants that provide more food, and in species of fish.

Ohno's hypothesis suggests that the two duplications happened in vertebrates within a short time, just as they started off on a new evolutionary branch. It might even have been the cause of their divergence, by creating a huge number of new genes that could undergo mutations and acquire new functions. The hypothesis has not yet been proved because some of the "tracks" have been erased. Most scientists are convinced that one genome duplication did happen; there is more debate about the second. Rather than a single event, there may have been several smaller ones in which large portions of DNA were copied. It may be possible to prove or disprove the hypothesis when complete genomes are available for a wider range of animals.

PHYLOGENY: UPDATING THE "TREE OF LIFE"

Scientists' new ability to read the genetic code has changed the way they arrange species into evolutionary trees. Linnaeus, Geoffroy Saint-Hilaire, Darwin, and the other great catalogers and classifiers of species had to draw their conclusions from plant and animal anatomy rather than an analysis of the genetic code. But painstaking work allowed them to go far beyond superficial appearances. They discovered, for example, that dolphins and whales were much more closely related to land animals than to fish. Whales and fish look similar until one examines their skeletons and discovers that whale fins are basically hands, covered with flippers.

Scientists call hands and whale fins *homologous* because they arose from a common ancestor. Homologies can also appear within species; the testicles of males and the ovaries of females are also homologous, because they arose from a single organ in an ancient ancestor. The antennae and legs of insects are homologous because they come from a common structure (in fact, a mutation in a single gene can cause legs to grow in place of antennae on the insect's head).

Scientists also discovered cases of *homoplastic* characteristics, features which look similar but evolved independently. Fangs have evolved many times. Snakes and cats did not get them from the same ancestor, and even within the cat family, saber teeth developed several times. Wings evolved separately in birds and bats. Root systems arose in plants at least twice; leaves developed at least three times.

Single molecules also undergo homoplastic changes. The shape of a protein determines its function, just as a part needs a certain shape to function in a machine. So unrelated molecules that perform comparable jobs have sometimes acquired very similar shapes through natural selection.

Darwin understood that deep similarities in anatomy revealed the evolutionary relationships between species, which meant that classification systems could be turned into *phylogenies,* or branching diagrams showing the evolutionary relationships between species. Ideally, all living creatures would be placed onto a single tree, spreading upward from the last common universal ancestor. The German scientist Ernst Haeckel (1834–1919) tried to draw one for the plant and animal kingdoms but could only go so far. He was unable to dig deeper into the past, to trace the evolution of single cells; that would require a look at their chemistry and genes.

Willi Henning (1913–76) fundamentally changed how phylogenies were made. In the 1950s and 1960s he introduced the term *clade,* meaning any branch of the tree with a common ancestor and all of its subbranches. Every organism belongs to several sets of clades, depending on what the diagram is intended to show. For example, the genealogy of a human family is a clade; the primates are another; the eukaryotes yet another.

Gene sequences have now become a main method for grouping organisms into clades. Phylogenetic trees have been built by comparing the DNA sequences of single genes from different species. This approach is not perfect, however, because of a phenomenon called *horizontal gene transfer* (HGT). Bacteria sometimes swap genes with each other, or "kidnap" them from an organism they have infected; viruses may do so as well. Sometimes they implant this information into the genome of another organism, which can confuse evolutionary trees—the way a person's family tree would have mistakes if one did not know that an ancestor had been adopted. Thus to build the most accurate phylogeny, many genes should be used—when possible, the entire genome.

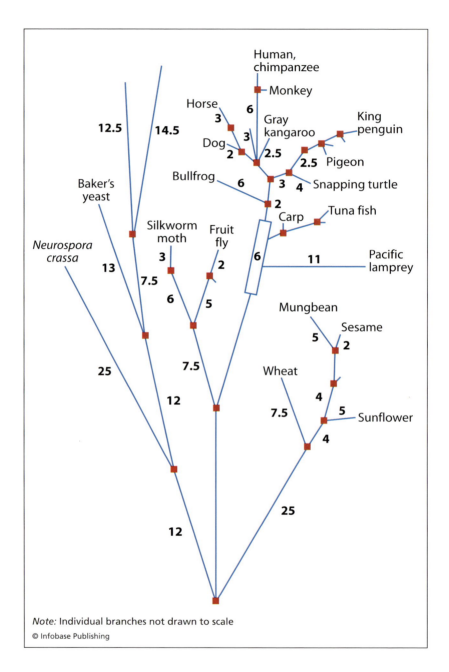

Human,
chimpanzee
Monkey
Horse
6
12.5
14.5
3
Gray
kangaroo
King
penguin
Dog
3
2
2.5
2.5
Pigeon
Bullfrog
6
3
4
Snapping turtle
Baker's
yeast
2
Tuna fish
Carp
Neurospora
crassa
Silkworm
moth
Fruit
fly
6
11
Pacific
lamprey
13
3
2
7.5
6
5
Mungbean
Sesame
5
2
7.5
Wheat
4
12
7.5
5
Sunflower
4
25
25
12

Note: Individual branches not drawn to scale

© Infobase Publishing

In 2006, Peer Bork's laboratory at the European Molecular Biology Laboratory in Germany began building a new tree
based on complete genome sequences. They combed data from

(opposite page) A "tree of life" can be assembled even from the spelling of a single gene. This tree was drawn using the DNA sequence of cytochrome c. It closely resembles trees made using more data, revealing that evolution and natural selection have left their "footprints" even on individual genes.

191 species and came up with 36 genes that could be compared because they had clearly shared common ancestors. The scientists discarded several of these because of hints that they might be examples of HGTs. The remaining information was combined into a super-tree which allowed scientists to carry out the most thorough comparison ever of the three domains of life—bacteria, archaea, and eukaryotes.

The results clear up some old controversies regarding the very early evolution of animals. Some trees in the past proposed that the vertebrates (which include humans) split off from another branch, which later split into separate clades for worms and insects. The new version groups things differently. Vertebrates and insects departed together from the worms, and diverged from each other later.

The new tree also provides a glimpse of much earlier events in evolution. Of today's forms of life, archaea appeared earliest, followed by bacteria. Bork's study showed that the earliest bacteria had genes that are crucial to survival under very hot conditions, so like many archaea, they probably lived in a hot environment, such as near a thermal vent in the ocean floor. A more recent study of proteins by Omjoy Ganesh, described in chapter 5, has confirmed this finding.

MINIMAL GENOMES AND THE LAST UNIVERSAL COMMON ANCESTOR

The kingdoms of archaea, eukaryotes, and bacteria likely inherited their most basic biological processes, such as storing hereditary material in the form of DNA, from their common ancestor. Looking for genes that are common to all forms of life has revealed some of the characteristics of the Last Universal

Common Ancestor. Another approach to investigating the features of LUCA is to try to list the smallest set of genes necessary for life—"the minimal genome." Some biological processes are so essential that every organism alive today needs them, which suggests that the first living cell must have possessed them.

Studies of this type have focused on genomes that are small to begin with, such as that of the bacterium *Mycoplasma genitalium*. Even though this cell has only 470 genes, some of them seem to be superfluous. Scientists have been able to remove 133 of these genes (in separate studies, not all at once) without fatal damage to the bacterium. As discussed earlier in the chapter, this process of paring down has also occurred in nature. *Mycobacterium leprae,* the bacterium that causes leprosy, is closely related to another disease-causing organism, the bacterium *M. tuberculosis.* But *M. leprae* has downsized, losing about 2,000 genes, and it is still able to survive.

LUCA was probably more complex than this minimal genome. A 2006 study by Christos Ouzounis' laboratory at the European Bioinformatics Institute in Great Britain scanned complete genomes and found 669 common "ortholog families," or groups of equivalent genes, which are known to participate in 561 biological functions. The number is higher than many scientists expected and shows that the ancestor must have carried out biological functions which we consider to be quite complex. Predictably these include copying, repairing, and altering DNA; transcribing genes into RNA and then translating that into proteins; cellular processes such as signaling and protection from extreme changes in temperature; well-developed membranes and mechanisms for shuttling molecules between them; and a sophisticated set of metabolic enzymes that can produce the building blocks of DNA, proteins, and membranes.

One surprise was to find molecules that snatch charged particles from the environment. Electrons are needed as energy, to drive chemical reactions, and perform other functions within cells. In the modern ecosphere, oxygen is a key source of these electrons, and some components of the ancient system seem to be suited to working in an environment with oxygen. Yet oxygen in the Earth's atmosphere is thought to have evolved late, after the activity of microorganisms had significantly changed the environment. "Oxygen-breathing" organisms could only survive once that had happened.

A possible explanation comes from a 2006 study by Jennifer Eigenbrode of Penn State University and Katherine Freeman of the Carnegie Institution in Washington, D.C. Their work suggests that there might have been more oxygen on the early Earth than previously believed. "Our evidence points to the likelihood that Earth was peppered with small 'oases' of shallow-water, oxygen-producing, photosynthetic microbes around 2.7 billion years ago," Eigenbrode states. "Over time these oases must have expanded, eventually enriching the atmosphere with oxygen. Our data record this transition."

No one knows exactly when LUCA lived, or how life developed before it arose, because all of the information available to scientists comes from the descendants of LUCA. It was clearly not the earliest form of life on Earth. A long process of evolution was necessary to assemble its complex biology. Once in place, those processes were obviously very successful; they have withstood the test of billions of years of evolution, in an enormous variety of creatures.

THE ORIGINS OF GENETIC DISEASES

Studies of the genomes of humans and other species are also revealing new connections between genes and disease. Most genetic diseases begin as a mutation in a single person and are passed along to his or her descendants, so they often occur at a much higher rate in particular regions of the world or specific ethnic groups. This makes it possible to discover when and where particular diseases originated; in some cases, the founder can even be identified by name. For example, a form of "protein C deficiency," which causes excessive, dangerous blood clotting, has been traced back to a marriage that took place in 1757.

Several thousand diseases have been linked to mutations in single genes. While each one is relatively rare, in total about 5 percent of the population is affected by such conditions. It

From Sequencing Organisms to Sequencing the World

How much life is there in a drop of seawater? In an ounce of farm soil? Until now, only a fraction of the Earth's species have been glimpsed. Microorganisms need to be brought into the laboratory to be identified and studied, but most do not survive in the lab. That problem will not be easily solved, but rapid new DNA sequencing techniques are allowing scientists to go into the field to capture their first look at the Earth's true diversity.

In 1985, Norm Pace, of the University of Indiana, had the idea that sequencing did not have to restrict itself to known organisms. Samples of water or soil could be taken and all the DNA found there could be analyzed to get a glimpse of the true complexity of life. Instead of revealing the sequence of a particular species, such metagenomic studies are more like the genome of a GPS coordinate. They uncover fragmentary traces of millions of genes without knowing precisely what organisms they come from.

This type of work is important because microorganisms play a crucial role in all food chains and are essential to the survival of every ecosphere, and metagenomics may be the only way to detect the presence of most species. To accurately measure the impact of humans on the environment—for example, to study the effects of pollution, global warming, or the effects of genetically-modified crops—there will need to be before-and-after surveys of the microscopic world. While events at this level are usually invisible, they may spell the difference between life and death for many more complex species, including our own.

In 2002, Mya Breitbart and Forest Rohwer, of San Diego State University, began taking an in-depth look at the ocean using a metagenomics approach. They discovered that 200 liters of seawater contain more than 5,000 species of viruses, and one kilogram of marine sediments may contain up to a million species. Samples of human feces contain over a thousand species. Hardly any of these had been seen before, and viruses were not the only organisms being found. When Craig Venter analyzed DNA taken from the Sargasso Sea, he found 148 new species of bacteria.

Craig Venter (*Liza Gross*)

Metagenomic studies also provide a new look at the effects of natural selection. Different environments are unique. Each one is more advantageous to some types of biochemical processes than others. A study of farm soil carried out by Peer Bork's laboratory revealed dozens of molecules involved in breaking down plant material; none of these were found in whale bones, taken from the ocean floor, or samples from the Sargasso Sea. Soil also contains many genes involved in how species respond to each other, for example, by the production of antibiotics. The Sargasso Sea yielded hundreds of genes similar to *bacteriorhodopsin*, a pigment that responds to light and allows cells to snatch energy from the environment.

A typical metagenomics study reveals thousands of types of genes that have never been seen in laboratory species. Some represent biological processes that do not occur in our cells. Investigating what these molecules do in exotic species of microorganisms will keep researchers busy for years to come.

is tempting to think that mutations which have a negative effect on a person's health should be wiped out quickly through natural selection. This is probably the reason for most miscarriages—but even dangerous mutations can survive by being recessive traits. If a person carries both a healthy and a damaged copy of a gene, there are often no symptoms, and the person may not be aware that he or she is a carrier. In other cases, people with one copy of a mutation will have a mild version of a disease, whereas two copies lead to a very dangerous, full-blown version. This is the situation with mutations of the globin gene that cause *sickle-cell anemia.*

Even these mutations have survived because what causes a disease at one period in a person's life may be beneficial during another. Mutations in globin and several other genes offer protection from malaria, an epidemic that has dogged humans throughout history, so these alleles have become extremely common in populations in tropical climates. People who inherit two defective copies of a gene in a related disease, called alpha-thalassemia, are 10 times less likely to suffer from malaria. Thus in one region in southern Nepal, 80 percent of the population carries the gene. The only known explanation for these phenomena is natural selection.

Genetic diseases caused by single genes follow Mendelian patterns of inheritance: If a person has one copy of the gene, half of his or her children will likely inherit the mutation, and they will pass it along to half of theirs. When two carriers of single copies marry, half of their children will have one defective copy of the gene, and a quarter of them will have two copies. If a damaged gene survives and is recessive, it may go unnoticed for many generations, until it has spread through the population and two people who carry single copies have children together. In earlier times, when people lived in the same small communities for centuries, this happened more often than it does today.

The presence of a mutation in a gene does not always lead to severe disease symptoms. Cells have sophisticated mechanisms to destroy RNAs or proteins made from some types of defective genes. These mechanisms may work better in some individuals than others. Thus in most cases, it is still impossible to say

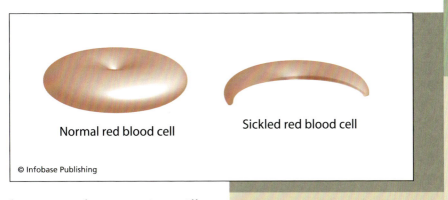

Normal red blood cell   Sickled red blood cell

© Infobase Publishing

how strongly a mutation will affect someone's life by looking at a gene. In other cases a second gene may "rescue" the first—in other words, it may fix the damage.

A damaged, recessive allele of hemoglobin causes red blood cells (right) to take on a strange "sickle" shape (left). Inheriting two copies of the gene usually leads to severe illness and is often fatal. People with one copy of the gene rarely show symptoms, but have a much lower risk of malaria infection.

Another reason that some disease-causing mutations survive is that they only become obvious or harmful late in life, after a person has had children and passed along the gene. As better living conditions and modern medicine have lengthened people's lives, scientists have discovered an increasing number of genetic diseases that strike in old age, particularly degenerative diseases. One form of Alzheimer's disease can be inherited because it is caused by a defect in a process that cuts a protein into small fragments. These clump together and form fibers that interrupt communication between nerve cells, which eventually die. But it takes many years for problems to become serious; this usually happens after people have had children, and until that time they are usually not even aware that there is a problem. Additionally, the development of many late-onset genetic diseases may be strongly influenced by the environment and a person's behavior.

Other problems that appear in old age may have been brewing as recessive genes in the population for a long time, going unnoticed because most people throughout history did not live long enough for symptoms to appear. The most common genetic disease in the Western world, beta-thalassemia, probably fits into this category. By disrupting the body's ability to absorb iron from food, it can cause serious problems, but they rarely develop before the age of 40.

Evidence suggests that most of the current inhabitants of Finland descend from a small population that had the gene that causes cystic fibrosis, a disease of the glands in which organs produce too much mucus and the immune system is deficient. The mutation can also be found at a very high rate among a small group called the Basques, living in a border region between France and Spain. Many Europeans carry this gene as well. Africans and their descendants have a high rate of sickle-cell anemia. Some cancers have been linked to an Asian heritage. A type of inheritable deafness called Usher syndrome began in a population in northern Israel.

Another interesting case can be found among the Amish in Lancaster County, Pennsylvania. In the 1740s, a couple immigrated to this area and left many descendants. The ancestor carried a mutation in the EVC gene. Over the past 270 years the group has maintained a strong religious and community identity, creating a small population that is somewhat isolated, so in the meantime descendants of the original couple have married. In the Pennsylvania group, 13 percent of the population carries a variant of a particular gene that is normally extremely rare. Someone who inherits two copies of this gene will suffer from the very serious Ellis–van Creveld syndrome. The disease gets its name from the two doctors who discovered it, Richard W. B. Ellis (1902–66) and Simon van Creveld (1895–1971), who met by chance on a train in England. In a casual conversation they discovered they were treating patients with the same symptoms, which included extra fingers and heart defects.

The same conditions that raised the frequency of this gene undoubtedly caused other unique features of the family to spread through the population, probably at a similar frequency. They have remained "invisible" because they do not lead to illness. If the Amish were to become separated from the rest of human populations for long periods of time, those features, plus genetic drift and new mutations, might eventually make the descendants of this family into a new species! But this would likely take tens of thousands of years of true isolation in which the community never mated with outsiders.

Advances in genetics and genome studies have helped researchers discover thousands of single genes that play a role in disease. Many more syndromes have been found to be the result of combinations of genes, and identifying them will require studies of huge populations. For many years this type of study has been carried out in Iceland by the company deCODE. The Icelandic population seemed ideal for this type of work because it is small and has been isolated for most of its history. Genetic studies require an accurate record of family relationships, and in Iceland nearly everyone can trace his or her family tree back 1,200 years. Additionally, public-health services have kept detailed records on the health of individuals for nearly a century. Access to those records was essential to the project, but normally they are considered private, and there was no way to obtain permission from the entire population. A special act of the Icelandic Parliament gave the company access to the information, but a few years later the country's Supreme Court overturned the decision. The company plans to continue working on the problem of multi-gene diseases using another strategy.

COEVOLUTION

During his voyage on the *Beagle,* Darwin realized that every organism is deeply entangled in a network of connections to other organisms and the physical environment. It depends on the food it eats, predators, the territory in which it lives, the climate, the seasons, the colors that surround it, and every other aspect of its surroundings. The theory of evolution created a very sophisticated view of these relationships. Because living things form part of each other's environments, they become a force of mutual natural selection, and coevolution refers to the way changes in one organism shape the others around it. As Darwin wrote in *On the Origin of Species,* "It is interesting to contemplate an entangled bank, clothed with many plants of many kinds, with birds singing on the bushes, with various insects flitting about, and with worms crawling through the damp earth, and to reflect that these elaborately constructed forms, so different from each other, and dependent on each other in so complex a manner, have all been produced by laws acting around us."

This interdependence is most obvious in the relationship between parasites and the plants and animals they infect, often called an evolutionary "arms race." Suppose that a horrible, fatal epidemic strikes. If some people happen to have partial immunity, they will survive to pass along their genes. If this happens for a long time, the entire population may become resistant and the disease may die out—unless some of the parasites have genes enabling them to escape destruction. In that case they can go on to infect new victims, meaning that a new cycle of disease may spread through the population—until it is stopped once again by immunity or medicines.

Parasites such as *Plasmodium,* which causes malaria, have recently evolved resistance to the most common drugs used to combat them. Malaria, tuberculosis, and several other diseases have been mankind's companion for thousands of years, possibly since the origin of modern humans, and genes that help people defeat them have survived. If scientists do not manage to completely eradicate AIDS, the same phenomenon is likely to occur with HIV. The disease has spread to so many people in Africa that it may eventually have a significant impact on human evolution, particularly if anyone has a natural resistance to it.

HIV itself demonstrates evolution in action: The virus undergoes such frequent mutations that it takes on a slightly different form in every person that it infects. This allowed researchers to prove that a sailor named Arvid Noe was the first to bring the virus to Sweden, and that a dentist in Florida infected several of his patients. Noe is not the first officially recognized victim of AIDS. Scientists at Rockefeller University, in New York, found traces of the virus in tissues taken from an African man in 1959. The first victim in the United States may have been a St. Louis teenager who died in 1969, according to studies carried out by Robert Garry's laboratory at Tulane University, in New Orleans.

Other diseases with a genetic basis have influenced both human evolution and the course of history. For thousands of years native populations in Central and Southern America were isolated from Europe and Asia. When Spanish conquistadores arrived in the 16th century, they brought along diseases that killed millions and led to an explosion of civil war throughout the continent. Europeans had lived with these viruses for so long that they had achieved a sort of armed truce with them, or at least a partial resistance. Additionally, their immune

Jared Diamond (*The American Physiological Society*)

systems may have been inherently stronger. In his 1998 book *Guns, Germs, and Steel,* which won the Pulitzer Prize, biologist and "biogeographer" Jared Diamond (1937–) offers a possible explanation. The open landscape of Europe and Asia encouraged travel, transportation, and trade, which meant that people were constantly exposed to new diseases. Smaller populations living in isolation—which was encouraged by the North-South orientation of the Americas—probably encounter the same parasites over and over; their immune systems do not have to be prepared for new threats all the time.

Changes in parasites and their hosts are only one example of coevolution. Another is mimicry—such as when one butterfly's wings begin to look more and more like those of another species, perhaps one which tastes disgusting to predators. And changes in the physical characteristics of a plant can put pressure on organisms that depend on it. One famous case involves the Madagascar star, an orchid, and an insect called the hawkmoth that pollinates it. Darwin had noticed that bees pollinate red clover by reaching deeply into the plant with their tongues; this meant that the depth of the flower's floral tube, where the pollen is deposited, has to match the length of the bee's tongue. The Madagascar star has an astoundingly long floral tube—30 centimeters long. In *On the Origin of Species,* Darwin predicted that somewhere there must be a moth with a 30-centimeter tongue able to pollinate it. Thirty years later, scientists found such an insect. This seems like an impractical solution to an "engineering" problem; it seems like it would have been easier for the moth to have a shorter tongue, and the flower to have a shorter tube. In 1997, Lutz Wasserthal of the University of Erlangen-Nürnberg, in Germany, discovered what seems to be the reason—also an example of coevolution. It is in the best interests of hawkmoths to stay as far away from the flower as possible, because a hungry spider might be hiding there.

Coevolution works on the largest scales of life—between organisms and probably between groups of species—but also on the smallest. Deep within the cell, most proteins carry out their functions by docking onto other molecules, like fitting two puzzle pieces together. This means that their shapes have to be attuned to each other. A change in a molecule exerts pressure on its binding partners to adjust as well—just as if a key's shape changes, the lock should be changed as well. Since those partners usually bind to other proteins, a single mutation can cause a chain reaction of coevolution.

The process is also at work at higher levels inside the body. From the point of view of a single cell, the environment consists of neighboring cells and molecules and other substances in the bloodstream, the temperature of the body, and countless other factors. All of these come into play in selection. The cell may need to migrate to form a tissue. This behavior is usually guided by signals on the surface of other cells, and any change in signals would exert pressure for the cells to learn to read the new signposts. Otherwise they may head off in the wrong direction and build different tissues. New organs likely originated in this way.

The type of selection familiar to most people involves the whole organism. The arrival of new predators, competition for

Malaysian leaf insect in Taman Negara National Park *(Wikipedia)*

resources with other species, or changes in food or the climate can exert powerful pressures to coevolve. Over the past several thousand years, human beings have deliberately altered other species through agriculture, the breeding of animals, and clearing large areas of land to build settlements. As humans travel across the globe at a rapid pace, they carry along pets and parasites, sometimes deliberately, sometimes not. Increasingly, humans are forcing other species to coevolve. Until Darwin framed the theory of evolution, no one realized that this was happening.

Part of this is a natural process that is true of every species. But thanks to genetic engineering, scientists no longer have to wait for a species to accidentally acquire desired traits; increasingly, it is possible to make custom-designed organisms. These skills might be needed in case of extreme environmental changes—for example, to design plants that could survive under new conditions. No one knows whether this will fundamentally alter the way that evolution works.

Coevolution produces an entire network of adaptations which seem well-tuned to each other. Different species interact like the parts of a finely-built instrument. This situation is not the result of intelligent, purposeful design, but rather of combinations of selective pressures. These are changing all the time. It is impossible to predict where the next pressure will come from, or how organisms will respond. In some cases environmental conditions may change so rapidly or to such an extreme that even large populations or organisms will not have enough variety to survive. The result is extinction, which has been documented throughout the fossil record.

EVOLUTIONARY DEVELOPMENTAL BIOLOGY

Today evolution, genetic science, and embryology have been woven together into a new way of thinking about life. This field is known as Evo-Devo: evolutionary developmental biology. It is based on the fact that closely related organisms have closely related genes. These produce RNA and protein molecules that usually work in similar ways in the cells of different organisms. In the early stages of an animal's life, a main job of these molecules is to tell cells how to build a body. The more closely related two species are, the more similar this process is. The same principle explains why humans are built more like insects than sponges.

As soon as Darwin and Wallace made their ideas public, embryologists such as the German Ernst Haeckel began trying to understand development in terms of evolution. Like many other scientists of his day, he was fascinated with the formation of embryos, considered one of the greatest mysteries of life. Until the early 19th century, the only way to study this phenomenon was through dissections. Then dramatic improvements in microscopes revealed that animals were composed of cells that divided over and over to produce entire organisms. This changed the way people thought of embryos, and they began trying to understand how cells shaped themselves into complete living creatures.

Haeckel's work followed up on that of Karl Ernst von Baer (1792–1876), who had noticed a fascinating phenomenon: Many species that are quite different as adults bear a strong resemblance to each other during some of the phases they go through as embryos. Robert Chambers, who proposed an early "evolutionary" theory in his 1844 book *Vestiges of the Natural History of Creation,* summarized the idea this way: "Physiologists have observed that each animal passes, in the course of its germinal history, through a series of changes resembling the *permanent forms* of the various orders of animals inferior to it in the scale. . . . The frog, for some time after its birth, is a fish with external gills, and other organs fitting it for an aquatic life, all of which are changed as it advances to maturity, and becomes a land animal. . . . Nor is man himself exempt from this law. His first form is that which is permanent in the animalcule. His organization gradually passes through conditions generally resembling a fish, a reptile, a bird, and the lower mammalia, before it attains its specific maturity."

This added to the mystery: Why did the embryonic cells of a monkey always produce a monkey, rather than a human? Without a theory of heredity, which Mendel had not yet invented, there was no way to answer the question.

Haeckel believed that evolution offered an explanation, and he formulated a hypothesis called *recapitulation*: As an individual organism develops (*ontogeny*), it replays the evolutionary history of its species (*phylogeny*). All life began as a single cell; so does an individual. The earliest multicellular life forms were probably ball-shaped, with just a few different types of cells; a human embryo also goes through such a phase.

Von Baer and Haeckel found fascinating evidence for their claims. At one stage a human embryo develops structures like gill slits, which then disappear. This only made sense, Haeckel said, in light of the fact that the distant ancestors of mammals were fish. Haeckel drew images of embryos at various phases of development to show how similar the body plans of organisms are at different stages. This work has been criticized because his drawings tended to emphasize the similarities rather than the differences between embryos. But most modern

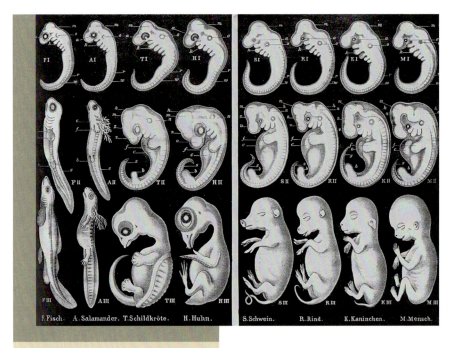

F.Fisch. A.Salamander. T.Schildkröte. H.Huhn.   S.Schwein.   R.Rind.   K.Kaninchen.   M.Mensch.

Haeckel's drawings *(Wikipedia)*

scholars believe that Haeckel was a victim of seeing what he wanted to see; he was not trying to perpetrate a fraud (which other scientists could have easily detected). In the days before photography, scientists had to draw what they saw under the microscope, which required a bit of interpretation. It was impossible to make a perfect sketch, particularly of tissues under the microscope.

Later scientists realized that "ontogeny recapitulates phylogeny" is a sort of side effect of the activity of genes rather than a law. It works best in organisms that evolved through a process that Stephen Jay Gould named terminal addition: species that arose by adding additional developmental phases to those of their ancestors (like adding boxcars to a train). Other species evolved in a different way: caterpillars, for example, do not go through the same type of larval phase seen in many other species. These species arose as mutations sent them off on different paths at early stages in their individual development, like trains sent off on parallel tracks early in their voyage. Haeckel's principle is not useful in such cases.

Haeckel did not have the technology to go beyond superficial descriptions of embryos; today scientists can look at the genes that guide an organism's development. This has ushered in a revival for part of the idea of recapitulation. The modern version states that embryos go through similar stages because the genetic programs that build complex organisms are similar. Different species inherited the genes that do this from a common ancestor. During part of their development they activate those genes in the same manner, which causes their tissues to develop in similar ways.

This is a new way of looking at such questions as why human embryos develop structures like gill slits. Far back in animal evolution, these features were created through a series of mutations that happened over a long period of time. Such accidents gave natural selection something new to work on, eventually shaping the structures into functional gills. Those have evolved further in fish and other animals. The reason they still appear in humans is likely that the processes that build them are useful for other reasons, and have therefore been maintained by natural selection.

The similar functions of most related genes in flies, mice, humans, and other species have allowed researchers to make effective models of human diseases. Mutations that cause problems in one species often cause similar problems in another. Thus a mutation in a gene called mdx in the mouse leads to muscular dystrophy, just as it does in humans. Other animal models mimic Down syndrome, epilepsy, and many more diseases. Such models are extremely important because researchers need to understand the mechanisms by which disease symptoms arise and challenge the body, and how they can be cured. This requires experiments that would be impossible or unethical to carry out on humans.

Another achievement of Evo-Devo has been to uncover very powerful genetic programs that have guided animal evolution for over a billion years. The *homeobox* (HOX) genes contain step-by-step instructions to create the head-to-tail organization of the body plans of organisms from worms to insects to human beings. HOX genes are arranged next to each other on a chromosome, like a string of words making up a sentence. As cells in the early embryos of animals divide, the HOX genes are

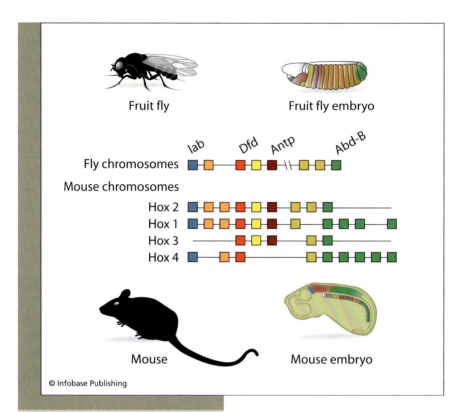

Fruit fly

Fruit fly embryo

Fly chromosomes

lab   Dfd   Antp   Abd-B

Mouse chromosomes

Hox 2
Hox 1
Hox 3
Hox 4

Mouse

Mouse embryo

© Infobase Publishing

As an embryo develops, HOX genes are switched on in a precise order to control the growth and formation of body structures in the head-to-rear direction. These genes have very similar functions and are activated in the same basic order in species of animals ranging from insects to humans.

activated, one after another, in cells, in the order in which they physically appear on the chromosome, building the head and lower segments of the body. This "sentence" has been preserved in all kinds of animals over nearly a billion years, which is extremely unusual; normally over the course of evolution, genes that act in sequences are split up and become part of new sentences. In this case not only have they remained together, but HOX genes are so important that the "meaning" of the sentence has stayed virtually the same.

In the 1990s, Detlev Arendt and Katharina Nübler-Jung of the University of Freiburg, in Germany, discovered another genetic program that creates the dorsal (back) and ventral (belly) regions of a huge number of animals. The pattern had been harder to find because early in evolution, in a remote, wormlike ancestor, it flipped over. Genes that build the back of an insect form the belly of vertebrates, and vice versa.

Other tissues and organs have been linked to key developmental genes, and this permits scientists to trace the very early evolutionary origins of complex structures such as the eye. These genetic programs were difficult to decode until Pat Brown of Stanford University and the California-based company Affymetrix created a new type of biotechnology called the microarray, or the *DNA chip*. The method uses probes made of DNA to detect when a cell uses a gene to create an RNA molecule. With probes taken from a complete genome, the entire activity of cells' genes can be watched. Thus, as cells differentiate, scientists can watch changes in their genetic programs.

These studies show that very similar programs are used in different organisms to create similar structures. They are also used to do different things in the same organism. The programs that create legs and antennae in insects, for example, are nearly identical. Thus, activating a particular gene in the wrong tissue can lead to the development of a leg—rather than an antenna—on an insect's head. Master control genes guide the creation of most body structures; thus a set of proteins created in a zone called the gray crescent of frogs launch the development of the head and brain. In 1924, Hans Spemann discovered that transplanting gray-crescent tissue into another newt led to embryos with two heads. There are many fail-safes built into cells and tissues to prevent the programming from going wrong, but if it does, the result may be serious flaws in development.

THE EVOLUTIONARY ORIGINS OF COMPLEX ORGANS

Darwin knew that it would be difficult to give a complete account of the origins of very complex organs like the eye. Its parts are so intricate and interdependent that it is difficult to imagine how it could arise through single mutations. But evolution predicted that once a single light-sensitive cell developed, natural selection could take care of the rest. It has recently become possible to test this hypothesis in eyes and other organs.

While the eyes of insects and mammals look very different, they are both based on light-sensitive cells called *photoreceptors.* With the invention of the electron microscope, scientists began investigating these cells and discovered that they come in two main types with different shapes. The two types share some common features: Each has a hairlike cilium that protrudes from the cell surface. Each has increased its sensitivity by evolving large surface areas, the way adding solar collectors to an array captures more sunlight and produces more power. But the surfaces have expanded in different ways in the two types. Ciliary photoreceptor cells have extensions on the cilium itself, as if branches have been plugged into an artificial Christmas tree. Rhabdomeric cells have added tiny protrusions on the surface of the main body of the cell, at the base of the cilium, like tall weeds growing at the base of a tree.

These cells looked so different that some researchers proposed that eyes must have evolved independently at least twice. But their conclusions were based on the shapes of the cells, rather than the genetic programs that guide eye development, and the latter tells a much different story. Each type of cell has a pigment molecule attached to a protein called an opsin. Receiving light changes the shape of the opsin, and this triggers chemical reactions in the cell.

In 2004 Detlev Arendt and Jochen Wittbrodt at the European Molecular Biology Laboratory, in Germany, looked deeper at the two types of photoreceptor cells and identified a pattern of genes—a sort of molecular fingerprint—that distinguished them from all other types of cells in the insect. Then they checked for related molecules in the genomes of mammals and fish. They also probed for these genes in the early larvae of a species called *Platynereis dumerilii,* which scientists consider a sort of "living fossil"—the species alive today, which is probably closest to the common ancestor of insects and vertebrates. A pattern emerged: In all of these different species, related genes

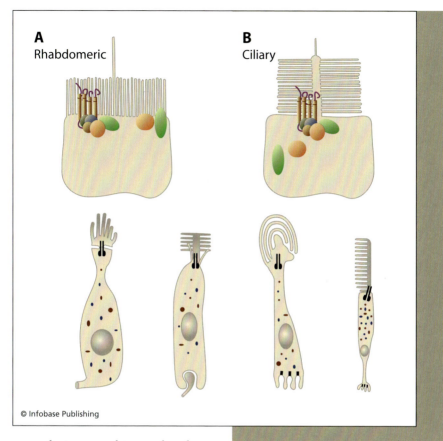

A
Rhabdomeric

B
Ciliary

© Infobase Publishing

were being used to make the two types of photoreceptor cells. So even though the eyes of today's insects and vertebrates look very different—even down to the architecture of the cells that compose them—they can be traced back to two types of light-sensitive cells, and those originated as a single type of cell.

The eyes of modern animal species are made of two basic types of cells called rhabdomeric and ciliary photoreceptor cells. Genetic studies of the two types show that many of the same genes are used to transform photons into signals that can be passed through the nervous system to the brain. This is strong evidence that eyes originated one time, in a common ancient ancestor.

Interestingly, many organisms use both types of photoreceptor cells, but in different ways. In some species they serve directly as eyes; in others, they evolved into retinal nerves that are used to transmit signals from the eye into the brain. It was another example of how nature has taken an existing structure (a light-sensitive cell) and spun it off into new versions with slightly different functions.

FROM GENES TO FORMS— UNDERSTANDING NATURE'S TOOLBOX

Studies of photoreceptors and other types of cells have answered some questions about the ancient origins of complex organs, but they have not explained everything. Ultimately, brains and eyes are built by single cells that blindly follow instructions encoded in their genes. How do mutations manage to make successful changes in such complex organs? Marc Kirschner of Harvard University and John Gerhart of the University of California, Berkeley, have developed a new way of thinking about the question. In a 2005 book called *The Plausibility of Life,* they introduced a theory called "facilitated variation." The principle reveals how nature has used the same basic elements within the cell to create an amazing variety of forms, making the "themes and variations" of evolution. While mutations and other changes in DNA are random, they rarely have completely chaotic effects. If an organism survives these changes long enough to be born it is because it has still managed to produce a body that functions well. The organism may not be as well-suited to its present environment as other members of its species; on the other hand, it may have better chances than others if external conditions change.

Even processes such as how well DNA is copied have been shaped by natural selection—if the system were perfect, there would have been no evolution and human beings would not exist. Cells separated by billions of years of evolution still share fundamental, core processes that can be traced back to the last universal common ancestor of life. Usually the differences between species stem not from throwing out these essential components, or creating entirely new ones, but from slightly changing when and where in the body they are active. For example, the same set of genes may be switched on at different times, or for longer amounts of time, or linked to other processes in new ways. A good analogy is a toolbox and an electronics kit that can be used to construct many different devices.

Occasionally, however, magnificent new tools have been added to the kit. Kirschner and Gerhart call them "core processes" and list several examples: the development of crucial biochemical reactions in the first archaea; the first eukaryotes (cells that store their DNA in a special compartment, the nucleus); the first multicellular organisms; the development of a left-right symmetrical body plan in animals; the rise of special neural crest cells that lead to the formation of the brain and spine; the development of limbs in the first vertebrates to live on land; and the development of the neocortex, a part of the brain responsible for perception and higher functions in mammals.

Most of these innovations are more than 500 million years old. The wide variety of life that has arisen since then has mostly come from changes in molecules that govern the details of these processes. There are two main reasons for the difference: Altering a core process is more likely to cause severe damage, and the genes for "regulators" are smaller. This means they are more susceptible to changes through mutations, the way that changing one letter in a four-letter word is usually more confusing than changing one letter of a 20-letter phrase.

Core processes have evolved special characteristics that give them flexibility—like Legos, they can be put together in an almost infinite variety of ways. They can also be linked together in new combinations that serve single-celled organisms or huge animals. These characteristics, Kirschner and Gerhart say, include weak linkage and *exploratory processes.*

Examples of weak linkage are very sophisticated programs in cells that always have a certain effect—for example, a particular type of chemical signal may always activate the same gene. What keeps these "circuits" from freezing up and only doing the same thing over and over, in all species, is the fact that they have multiple inputs; they can be stimulated in different ways. This happens in cells, but there are parallels in behavior. Suppose that every time a person is afraid, he runs. Weak linkage is like the fact that many different events might frighten him.

Exploratory processes are random events that are like small experiments whose only aim is to see what happens. For example, cells take small *tubulin* molecules and link them up into long tubes called *microtubules.* These stretch through the cell and are used as transport routes for molecules. They are constantly being built and taken apart, sprouting off in all random directions, which is useful because the cell cannot foresee exactly when and where they will be needed. To be used, they have to be stabilized, and this is the work of other molecules. So a microtubule keeps "exploring" until it receives the signal to stop and do something else. The cell can take advantage of this in different ways, depending on what the control molecules are up to. In the daily life of the cell the microtubule system functions as a protein motorway, aiming outward from the nucleus. When it comes time for cell division, the whole network is rebuilt. Now it aims inward, and is used to separate chromosomes into two daughter cells. The switchover is orchestrated by control molecules.

Kirschner and Gerhart see evolution as a whole working this way. An organism's genes are the product of evolution. Their design gives them certain functions, but there is weak linkage because the environment can provide an infinite number of inputs. Thus the brain has a structure formed by evolution, but the same brain is open to different kinds of input. These determine the language a person will speak, and whether she becomes a Roman of the year 50 C.E. or an American of the 23rd century; the same brain is capable of both.

Mutations also have an exploratory nature—they happen in random ways. Some are harmful, which leads to a breakdown of the organism (the way that unnecessary microtubules are dismantled). Others are successful and become stabilized by the environment—in other words, they permit an organism to pass along its DNA.

Weak linkage and exploratory behavior can be seen at every level of complexity in every biological system, from single molecules to the complex behavior of an animal. These principles can be read in our genes: The tubulin protein, used to make microtubules, has remained almost unchanged since the development of the first eukaryotic cells a billion years ago. But the regulatory molecules that control how it is used have changed dramatically, in different species, and in different types of cells.

FUTURE QUESTIONS AND REFLECTIONS

"Science consists in grouping facts so that general laws or conclusions may be drawn from them," Charles Darwin wrote. Before evolution, there was no systematic way to make sense of life and nature; scientists had to regard the living world as a collection of single, separate phenomena, each created by miraclous events. Religions assumed that nature functioned fundamentally differently long in the past than in the present day, so the origins of the world and of life were considered outside the realm of scientific investigation. But theologians disagreed about the nature of God and the universe, and their opinions did not quench human curiosity; as scientists carefully examined the world around them, they began to understand that processes they could observe had been going on for a long time. What if the universe had been obeying the same rules forever? This idea led to a new method of understanding the world that revolutionized astronomy, geology, and finally the life sciences.

A scientific theory is an interwoven set of hypotheses that are put forward with the expectation that they will be scrutinized very critically by laboratories across the world, by scientists from different cultures and belief systems. Real science is global and democratic: If an experiment is set up in a particular way, its results can be verified by any person from any religious background, any culture, any race, and any social class.

If a hypothesis does not stand the test of time, it must be discarded or changed, and this has happened many times in the history of science. One of the best things that can happen in a researcher's career is to make a discovery that successfully overturns a leading theory. This usually leads to fame, more freedom to pursue themes that the scientist is interested in, and sometimes a Nobel Prize. Every day over the past 150 years researchers have eagerly put evolution to the test, and it has withstood them all. At any time a paleontologist or a geneticist might have found something that contradicted the theory. But that has not happened. Instead, every new fossil that has been found, all the DNA sequences and other new types of data that have appeared, and every experiment that has sought to overturn Darwin's work has done the opposite, lending more support to the idea that the living world is governed by the principles of variation, heredity, and natural selection. When new layers of rocks are exposed, they confirm geologists' estimates of the age of the Earth. When paleontologists dig, the fossils they encounter conform to a logical, evolutionary sequence of events, and the same thing happens when a new genome sequence is decoded.

There have been many debates about the details. This is a healthy process, and it will continue because of the extreme complexity of life and its history. As important and useful as the theory has been, some questions about evolution will never be solved. Scientists will never have a complete, gapless record of the history of life from its beginnings to the present. Most fossils do not survive, which means that significant parts of history have been lost forever. We will probably never know exactly what happened during some of the most interesting eras in the development of life—for example, the first billion years of our planet's history, as the first cell arose from inorganic substances. Another gap hides the era in which the first animals arose, because they did not have hard body parts that usually leave fossils.

Nor will scientists obtain a gapless record of the mutations and changes in DNA that transformed past species into others. But studies of the HIV virus, laboratory animals, and domesticated plants and animals prove that this process is going on right now all around us, and comparing the genomes of modern species shows that it has happened throughout the past. No one has yet discovered a fossil, a living organism, or a biochemical process that appears to have been created spontaneously, an orphan without a parent. Science has solved the question of the chicken and the egg, and where "Adam and Eve" came from.

Darwin and Wallace could not answer many questions that their theory raised 150 years ago, such as the issue of how heredity works. In the meantime most of those questions have been answered or are on the verge of being answered. Scientists still do not completely understand how a single genome produces so many different forms. Muscle, blood, and nerve cells look incredibly different; anyone who found these cells floating alone would surely think they belonged to completely different species. Yet they are the product of the same collection of genes. In a similar way, each organism is different from its relatives. Somewhere in the tiny proportion of genes (less than 1 percent) which distinguish humans from all other animals lie the rules of freedom and constraint that make each person different and yet so much the same.

Understanding evolutionary processes will be essential to coping with the future, partly because of the influence that humans are having on the Earth. All organisms change their environments (an extreme example is the way ancient organisms completely modified the world by filling it with oxygen) but humans have been doing so consciously—first through breeding, now through genetic engineering. Humans are the first creatures in history to understand the forces that produced their own species. This has already begun to influence our evolution. Genetic engineering has created new crops that have been on the market for over a decade, and genes are being implanted into animals to improve their nutritional value or other qualities. Diseases that would have been fatal in the past—an element of natural selection—are being defeated through biotechnology. People who would have been victims now survive, and this will change the genes of future generations. Natural weaknesses in the human immune system are being strengthened and supplemented by drugs and genetic tools. By changing the organisms around us, we are changing our own environment, and ourselves. By making new tools, we are changing the nature of the toolmaker, and it is vital to understand this process so that we will be able to use this newfound power wisely.

Despite everything that it has accomplished, evolution is a theme that divides society, much more in the United States than in most other parts of the world. The fact that it did so 150 years ago is not surprising, given that the theory of evolution ran so contrary to old traditions that everyone had been taught about the origins and purpose of life. Evolution joined a long list of scientific discoveries that were upsetting when they were first announced to the public. But in today's society the debate between scientists and religious fundamentalists has created a very strange situation.

This is most obvious in repeated attempts to ban the teaching of evolution in schools or to require that the religious philosophy called "intelligent design" be given equal time in the classroom. This movement is a modern version of "natural theology," which arose at the end of the 18th century as a study of nature for the sole purpose of gathering evidence that a particular type of God exists. All observations are interpreted in such a way to support this assumption, and contradictions that arise are attributed to evil forces or the imperfection of human perception or the mind. People have been accused of evil for believing that the Earth orbits the Sun, for daring to observe that grasshoppers have six legs rather than four (as the book of Leviticus claims), and for claiming that rocks in deep strata in the Earth are older than those on the surface.

Today the movement has a different name; calling it "theology" would clearly demonstrate its religious intent. But a new name does not change the fact that it has nothing to do with science. The intelligent-design movement claims to be scientific because it examines nature and picks and chooses from scientific facts. However, there is a huge difference between real science and pseudoscience, even one that uses charts, numbers, and impressive language that sounds scientific to nonscientists. The most obvious sign is how the intelligent-design movement ignores or seeks to discredit any evidence that does not support its initial hypothesis, and this is not how science works.

Every day researchers discover things that they cannot explain. This does not mean that they have found evidence of miracles that will never be explained by science. Airplanes appear to defy the laws of gravity. Obviously they do not; there is a scientific explanation for their behavior. But if a proponent of intelligent design from a Stone Age culture ever saw an airplane, he surely would claim that it was a miracle. Since he could not explain it, he might even have held it up as proof the existence of the Judeo-Christian God and the account of creation given in the book of Genesis. By claiming that any other opinion was misguided or the product of evil, he might stop his friends from wondering how a machine could fly. And then there would be no reason to try to build one.

Just as natural theology has been renamed in order to try to bring religion into schools, it has been reinvented several times over the past 150 years in order to cope with each new type of evidence for evolution. A main challenge for 19th-century critics was to explain away "useless" features of animals that seemed related to "useful" features in other species. Why should pigs have extra toes that never touched the ground and seemed to have no other function? Other features of animals seemed so well-designed that they could never have evolved—such as the eye, or tiny bones in the inner ear that permit hearing. In the meantime, paleontologists have found evidence for transitional forms of nearly every organ in fossil plants and animals and their living descendants.

When scientists learned to read the genetic code, they discovered a parallel situation: Humans contained DNA sequences that were nearly identical to those of mouse genes, but the human form of the gene no longer functioned because it had undergone mutations. Suddenly proponents of intelligent design had to find a way to explain away such "pseudogenes" without referring to evolution. Computer scientists were hired to try to find an alternative explanation. It is possible to explain away anything by ignoring the rules of logic or calling it evil, but facts cannot be erased, and in the end they win out over theological opinions.

Michael Behe was introduced in chapter 4 as a biochemist, a proponent of intelligent design, and the author of *Darwin's Black Box*. In 1994 he claimed that the lack of "missing links" between whales and land-dwelling mammals posed a serious challenge to evolutionary theory; by the end of the same year three such fossils had been found. He stopped talking about whales and began focusing on the "irreducible complexity" of proteins that make up the flagella of bacteria—which could not, he claims, have evolved. Now that scientists have found forms of nearly all the proteins making up flagella in earlier forms of life, where they carry out different but useful functions, he will have to start looking elsewhere for challenges to evolutionary theory. At some point he may find something like an airplane, a fact that may take many generations to explain. There is a point to this: In fact, he is strengthening the case for evolution. Any living system that is difficult to understand in terms of the theory is a very good project for research, and scientists usually take it on eagerly. So far, all the results have been compatible with the theory.

Most advocates of intelligent design accept two pillars of evolutionary theory—heredity and variation. They know that children inherit genes from their parents and they do not claim that humans are all clones of each other. Most also recognize that mutations occur, a third element of evolution. Most also recognize that the environment plays a role in keeping populations under control, ensuring that the Earth has not been overrun by sea slugs, elephants, or birds.

Where, then, are the problems? The largest issues seem to be how the first living creatures evolved, whether natural selection can produce new species, and whether humans have been subject to natural selection in the past. There is so much evidence for the latter two points that they are no longer the subject of serious scientific debates. As to the origins of life, chemists have not yet been able to explain the steps that produced the first living cells from nonliving materials, but very few of them doubt that it could have happened. (Strictly speaking, this is not a part of evolutionary theory anyway.)

People also claim that evolutionary theory is threatening because it disproves the existence of a supernatural being. It does not; it says nothing about the issue, but it is true that Darwin and many other people have said that evolution influenced the process by which they became agnostics or atheists. On the other hand, many scientists and others who accept evolution as an explanation for the development of life on Earth continue to hold religious beliefs. They do not believe that the theory renders life meaningless—their view is shared by the Dalai Lama, the Pope, and other religious leaders around the world.

The real problem that many fundamentalist religious groups have with evolution probably stems from the fact that evolution is a materialist philosophy. Until 1859 there was no tenable scientific theory that could explain extinctions, the fossil record, or changes in the types of life on Earth. Evolution, biochemistry, and other life sciences provided a materialist explanation for some biological phenomena; in other words, they stated that it was not necessary to refer to supernatural forces to account for the existence of material phenomena such as living organisms. But evolution and molecular biology have not yet provided a complete explanation for how consciousness or spiritual experiences arise from the material brain, body, and environment. Many scientists, including Richard Dawkins, believe that in the future there will be materialist explanations for these phenomena, and that the entire universe will one day be explainable without referring to supernatural forces. They also criticize the methods by which religions link specific phenomena (for example, consciousness) to particular religious doctrines (such as Christianity). But the scientific community does not have a unified theory or set of beliefs about these issues.

The United States and many other countries have shown that it is possible to create a society in which people are free to believe whatever they like while still living according to ethical principles, like respecting the value of human life and the rights of others. If it is possible to live harmoniously in a society that permits different religions that have radically different views of creation and the nature of a God or gods, it must also be possible to live in a society where people are free to choose not to hold religious beliefs.

In a 2006 book called *Creation and Evolution,* Pope Benedict XVI stated that he was a believer in *theistic evolutionism,* a philosophy that holds that God used evolution to create life. The Pope has said that it is absurd to reject evolution in favor of faith or to reject God because of science, because it is possible to accept both. He has pleaded for people to be reasonable and to stop turning evolution into a polarizing issue. Most of the guiding theologians of our time—and most scientists—believe that human beings can continue to hold onto moral values and find meaning for their lives in a world produced by evolutionary laws. After all, humankind has been doing so since the origin of our species.

Chronology

sixth century B.C.E.	The Greek philosopher Anaximander claims that life arose in the sea; at some point fishlike creatures moved to land, where they gave birth to humans.
221 C.E.	Julius Africanus uses the Bible to calculate the age of the Earth, arriving at a date of ca. 4000 B.C.E. Over the following centuries, Bishop Eusebius of Caesaria and many other Christian scholars follow his example and arrive at similar dates.
1543	Nicolaus Copernicus's book *On the Revolutions of the Heavenly Spheres,* published the year he dies, proposes that the Earth orbits the Sun; it is harshly criticized by Martin Luther and other religious authorities.
1632	Galileo publishes *A Dialogue on the Two World Systems,* leading to his questioning by the Inquisition and a lifetime sentence of house arrest. In other writings, Galileo formulates many of the fundamental principles of modern science.
1650	James Ussher concludes from the Bible that Creation occurred on the eve of Sunday, October 23, 4004 B.C.E.
1669	Niels Stensen, working for the Medici family in Florence, proposes that strata in rock reflect

geological time, with the oldest layers at the bottom and newer layers above. He recognizes that fossils are remains of living creatures.

1687   Sir Isaac Newton's book *The Principia* describes the three laws of motion, including a description of gravity, called the first "natural law." Modern science is born as a mixture of empirical, experimental, and theoretical research.

1735   Carolus Linnaeus publishes the first edition of his *Systema Naturae,* the primary antecedent of the modern science of taxonomy. Linnaeus believed in an early concept of common descent, with all plants having evolved from a common ancestor, but humans and animals having been directly created by God.

James Hutton independently discovers the connection between strata and geological time. He proposes modern explanations for the origins of different types of rock and finds evidence of numerous cycles of upheaval and erosion of the Earth's crust.

1794–1796   Erasmus Darwin publishes *Zoonomia,* which contains a Lamarckian-like version of species change.

1798   Thomas Malthus publishes "An Essay on Population," a crucial influence on the concept of natural selection.

1800   William Smith begins a detailed map of the strata of rocks in Great Britain, discovering that each geological layer contains a unique collection of fossils. This shows that the species present on Earth have changed over time.

1802   Jean-Baptiste Lamarck publishes *Research on the Organization of Living Bodies,* in which he claims that species change through environmental influences and through their own activites and behavior. His philosophy of species change is based on the notion that animals seek to become more perfect and better adapted to their surroundings.

William Paley publishes a monumental work called *Natural Theology,* which examines natural phenomena in order to reveal God's design of the world.

1803   Erasmus Darwin's book *The Temple of Nature,* published posthumously, promotes a version of species change.

1813   Two essays written by William Wells, an English physician, are read to the Royal Society. They contain the first description of natural selection and claim that it has shaped the human species.

1817   Georges Cuvier publishes *The Animal Kingdom,* in which he argues that all the characteristics of a species are attuned to fit its lifestyle. His studies of fossils show that they were living creatures that have become extinct.

1820s   Étienne Geoffroy Saint-Hilaire's studies of animal anatomy point out unusual relationships between species and features that seem to have lost their functions over time.

1830   Charles Lyell publishes *Principles of Geology*.

1831   Charles Darwin departs on his voyage on the *Beagle*.

A book called *Naval Timber and Arboriculture,* by Patrick Matthew, describes a form of natural selection as a universal force that shapes living creatures.

1838   Darwin formulates the theory of natural selection.

1844   Darwin writes an essay on evolution by natural selection, locks it in a drawer, and begins a study of barnacles that will last nearly a decade.

1851   Henry Freke, an Irish physician, proposes that a type of natural selection has shaped species.

1854   Alfred Russel Wallace departs for the Malay Archipelago.

1856   The first remains of Neanderthal man are discovered by Johann Fuhlrott in the Neander Valley, in Germany.

1858   Papers by Charles Darwin and Alfred Russel Wallace regarding evolution by natural selection are read at a meeting of the Linnean Society in London. Neither man is present.

1859	Charles Darwin publishes *On the Origin of Species.* The complete first print sells out on the first day.
1860	Thomas H. Huxley and Samuel Wilberforce, Bishop of Oxford, debate evolution in a huge public meeting which sets the tone for a long confrontation between conservative religious sects and evolutionary science.
1865	Gregor Mendel publishes a paper on the principles of heredity, which will eventually become the foundation of the science of genetics. The work reveals how parents pass along traits to their offspring.
1869	Francis Galton, Charles Darwin's cousin, begins a series of experiments intended to prove Darwin's conception of how heredity works; they achieve exactly the opposite. Galton promotes positive eugenics as a means of improving humanity.
1874	Elisha Harris begins a long-term study of criminal behavior in an extended family in New York. Misinterpretations of his data fuel negative eugenics movements.
1890s	American physicians begin sterilizing women with a history of "psychological problems" based on a flawed concept of evolution.
1891	Eugene Dubois discovers the first *Homo erectus* fossils in Java, Indonesia.

1900 Hugo de Vries, Carl Correns, and Erich Tschermak von Seysenegg rediscover the work of Gregor Mendel, and publish papers on heredity the same year.

1902 Archibald Garrod discovers that a form of arthritis follows a Mendelian pattern of inheritance, which means that it is a genetic disease.

1907 The state of Indiana passes a law under which people who were judged to be insane or "hereditary criminals" were sterilized. Within two decades, 30 states had passed such laws.

1908 Godfrey H. Hardy & Wilhelm Weinberg create the Hardy-Weinberg principle, a mathematical formula which describes the movement of genes through a population.

1909 William Bateson coins the term "genetics."

Thomas Hunt Morgan discovers the first mutations in fruit flies, *Drosophila melanogaster,* bred in the laboratory. This leads to the discovery of hundreds of new genes over the next decades.

1924 Raymond Dart discovers the first hominid fossil from Africa, *Australopithecus africanus.*

1925 John Scopes, a Tennessee high school teacher, is arrested and fined for teaching evolution in the classroom.

1931 Ronald Fisher publishes *The Genetical Theory of Natural Selection,* a mathematical model which shows that natural selection can account for the movement of genes through a population.

1932   Adolf Hitler becomes Chancellor of Germany and passes a law that blind, deaf, mentally ill, and "malformed" people are to receive mandatory sterilization. These policies become increasingly severe, are expanded to include Jews, political dissidents, and other "undesirables," and eventually result in the Holocaust.

1937   Theodosius Dobzhansky publishes *Genetics and the Origin of Species,* a milestone in linking the fields of evolution, genetics, and paleontology. Dobzhansky and his contemporaries achieve the "Modern Synthesis" of these fields.

1953   Francis Crick and James Watson publish a paper which describes the structure of DNA. Their model proves that genes are made of DNA and suggests how mutations can arise. This launches the field of molecular biology which shows, over the next 20 years, how the information in genes is used to build organisms.

1968   The U.S. Supreme Court declares laws against the teaching of evolution in schools unconstitutional.

1972   Stephen Jay Gould and Niles Eldredge propose the concept of punctuated equilibrium, suggesting that evolution does not always proceed at a slow, even pace.

1972	Alick Walker proposes that birds and crocodiles descend from a common ancestor.
1973	John Maynard Smith uses a type of mathematics called game theory to investigate various aspects of evolution.
1977	Walter Gilbert and Frederick Sanger develop a new method to decode the DNA sequences of organisms. The rapid development of this technology eventually leads to the human genome project and the completion of the genomes of hundreds of organisms.
1980	Luis Alvarez and other scientists discover evidence of an asteroid which struck the Earth 65 million years ago and propose that this caused the extinction of the dinosaurs and many other species.
1983	Walter Gehring's laboratory in Basel, Switzerland, and Matthew Scott and Amy Weiner, working at the University of Indiana, independently discover HOX genes: master patterning molecules for the creation of the head-to-tail axis in animals as diverse as flies and humans.
1985	Norm Pace proposes the idea of metagenomics: sequencing all the genes in a particular environment rather than only looking at single species.
1995	Walter Gehring's laboratory shows that a gene called Pax-6 triggers the development of

eyes in insects and vertebrates, evidence that very different types of eyes had a common evolutionary origin.

1997   Svaante Pääbo and colleagues at the University of Munich obtain and analyze Neanderthal DNA, demonstrating that the hominid species was not a direct ancestor of modern humans.

1999   The Kansas state school board removes teaching of evolution and cosmology from required high-school science curriculum and testing.

2000   Scientists complete a "working draft" of the human genome. The complete genome is published in 2003.

2006   Pope Benedict XVI states his personal belief in theistic evolutionism and calls for people to stop using evolution as an issue that polarizes science and religion.

2008   A French team headed by Michel Brunet establishes that a fossil skull found in Africa belongs to the earliest known hominid closely related to modern humans and chimpanzees, dated at ca. 7 M.Y.A.

Glossary

adaptation a result of natural selection acting on the differences in individuals of a species. If an environment stays basically the same over long periods of time, this process generally increases organisms' fitness; in other words, it improves their chances of surviving long enough to have offspring.

alkaptonuria a hereditary form of arthritis, the first human disease found to follow Gregor Mendel's laws of heredity after the rediscovery of Mendel's work at the beginning of the 20th century. This showed that the disease was due to a flaw in a recessive gene.

allele one variant of a single gene. Humans usually have two copies of each gene, one inherited from each parent, located at the same positions in their two chromosomes. These may be identical or different alleles.

altruism behavior that benefits another organism at a cost to oneself. It is usually considered to be the opposite of selfish behavior.

anatomical isolation a situation in which two members of a species no longer mate for reasons having to do with the physical makeup of their bodies. This can separate two subpopulations and eventually create two species out of one.

anther the male, pollen-bearing structure in plants

archaea single-celled organisms that are thought to be the oldest types of cells on Earth, ancestors of bacteria and eukaryotes. Many are extremophiles, living at high temperatures or in other extreme environments.

archetype the most standard representative of a species or group of organisms. In a wide range of similar species, the archetype is often the most similar to the ancestral species.

bacteria   single-celled organisms without a nucleus, which can be found everywhere on Earth. Bacteria make up one of the three most basic domains of life (the others are archaea and eukaryotes).

bacteriorhodopsin   a pigment molecule in archaea that transforms light into energy by absorbing photons and pumping protons into the cell. Its structure and function are similar to light-sensing molecules in animal eyes.

base pair   a unit made of two DNA nucleotides, either an adenosine bound to a thymine, or a guanine bound to a cystine

biometrics   the study of any physical characteristics of organisms that can be measured, such as weight or height

bottleneck   an event in which only a few members of a species survive. This usually happens in extreme circumstances such as natural catastrophes; natural selection or chance determines which organisms survive. As a result, the genes of only these organisms (possibly even genes that lead to disease) are passed down to future generations.

Cambrian explosion   an evolutionary phenomenon that occurred during the Cambrian era of geologic time, which lasted from about 540 to 500 million years ago. This period saw the development of a huge variety of new forms of animal bodies, including most of the main types that populate the Earth today.

capillary electrophoresis   a method that separates molecules such as DNA into different groups based on the fact that they carry different charges

catastrophe theory (catastrophism)   an idea proposed in the 18th and early 19th centuries to try to explain dramatic historical events such as the rise of mountain ranges or extinctions. The theory proposed that at some periods in history, the Earth was subject to far more extreme forces than those at work today.

chromosome   large, compressed clusters of DNA and many other molecules found in the cell nucleus

clade   a method of grouping individuals, species, or groups of species by clustering ancestors and their descendants into a tree diagram that shows their relationships

cloning   an asexual form of reproduction in which all of the genes needed to make up a new organism come from the mother. Unless mutations occur, the offspring will be an identical genetic copy of the mother.

competition   a state in which some individuals have better chances to survive and reproduce than other members of their species

creationism   in Judeo-Christian traditions, the belief that species were created by God in the way described in Genesis, the first book of the Bible, rather than through a long process of evolution

discontinuity   a concept developed by William Bateson to explain mutations that cause very dramatic, sudden changes in an organism. Because it was not understood that mutations in genes can cause both small and large changes in an organism's body, he regarded this as the probable cause of the development of new species, rather than the gradualist views of Darwin.

DNA (deoxyribonucleic acid)   the molecule that contains the hereditary information of all species. DNA encodes RNA molecules, many of which are used to produce proteins.

DNA fingerprinting   a method of comparing two DNA samples to determine whether they come from the same person or from people who are related to each other

dominance   in a situation where an organism has different alleles of the same gene, the dominant allele is the one that influences what happens in cells and the body, and a recessive one does not. Having a single copy of one dominant allele has the same effects as having two copies of it.

eugenics   deliberate attempts to change the human species by encouraging specific types of people to breed (positive eugenics) or by preventing types from doing so (negative eugenics)

eukaryotes   one of the three major domains of life (the others are archaea and bacteria). Eukaryotic cells are unique because their DNA is kept in a subcompartment, the nucleus. All plants and animals, as well as yeast and many other unicellular organisms, are eukaryotes.

evolution   the scientific theory first stated by Darwin and Wallace that explains the origins of new species from existing ones based on the principles of variation, heredity, and natural selection.

evolutionarily stable strategy (ESS)   a type of behavior that results in long-term stability in the genome of a species. The term was invented by game theorists to analyze the short- and long-term effects of behaviors such as competing for mates and territorialism.

evolutionary psychology   a way of studying human behavior and cognition from the point of view of evolution. Evolutionary psychologists try to link the activity of the human brain today to natural selection that occurred in our species' history.

exploratory processes   random events that arise from normal processes in cells or organisms' bodies. Sometimes these events prove to be helpful to an organism and are then "solidified" by natural selection, becoming an important part of the organism's normal functions.

extended phenotype   aspects of an organism's life such as language, social behavior, or the objects it makes that are not purely physical traits, but which may influence the way natural selection works on them

fitness   a measure of how well an organism with a particular set of genes is able to survive and reproduce under specific environmental conditions

fitness landscape   a metaphor using the image of a map to chart the fitness of members of a species. Peaks and valleys show how many organisms have a high degree versus a low degree of fitness.

fossil   any remaining trace of an organism that lived in the past. Usually this refers to mineralized body parts, but it can also refer to imprints or other indirect evidence of an organism's existence.

founder effect   a loss of variety in a species or group that occurs when only a small number of individuals survive to pass along their genes, and future generations inherit only a small subset of the variety that used to exist in a species. The fact that very few buffaloes survived extinction, for example, means that a founder effect can be seen in their descendants.

game theory   a mathematical analysis of the behavior of evolutionary "players," such as the members of a species, genes, or other biological units. The goal is to predict what the outcome will be when players behave and interact in different ways.

gemmule   a hypothetical "particle of inheritance," usually thought to exist in body fluids, that was thought to transfer hereditary information between parents and their offspring before the discovery of genes

gene   a sequence of nucleotides in a DNA molecule that holds the information needed by a cell to create a protein

gene duplication   a mistake that frequently happens when DNA is copied, producing one or more new extra copies of a gene

gene pool   the complete set of genes found in a group or species at a specific time

genetic drift   a change in a species' collection of alleles or genes when the species is not under significant pressure from natural selection

genetics   the study of genes and how they influence processes in organisms

genome   the complete set of DNA in an organism. The term is often used to refer to a "representative" set of genes from a species, as in "the human genome," although each individual's genome is slightly different.

gradualism   the theory that the physical and chemical processes that shape the world today are the same as those working in the distant past. This is in contrast to catastrophe theory.

Great Chain of Being   a philosophical movement developed by the ancient Greeks which holds that living beings exist in a continuous gradient of forms, from simple to complex, from imperfect to perfect

Hardy-Weinberg law   a mathematical formula showing that the frequency of dominant and recessive alleles in a population should stay the same from generation to generation, unless natural selection or some other type of bias is at work

heredity   the process by which organisms pass on physical characteristics to their offspring

heterozygote   an organism that has two different alleles for the same gene

homeobox (HOX) genes   genes that contain a code called a "homeobox," which gives them the ability to activate other genes. A subset of these genes plays a crucial role in building the head-to-tail body structure of embryos. They appeared before the evolution of multicellular life and have been passed down to nearly all animals whose bodies have left-right symmetry.

hominid   a member of the family of great apes, including humans, chimpanzees, gorillas, orangutans, and their extinct relatives

homology   a similarity between two organisms—in their genes, cells, organs, or bodies—arising from the fact that the features have been inherited from a common ancestor. Even two structures in a single organism are called homologous (for example, legs and antennae in insects) if they arose from one tissue in an ancestor.

homoplasy   characteristics of organisms that have similar structures or functions but which did not arise from a single ancestor. For example, the wings of birds and bats are similar, but they evolved along two different routes.

homozygote an organism with two copies of the same allele for a particular gene

horizontal gene transfer (HGT) a process by which an organism's genome acquires genes from a cell that is not its parent. Bacteria, for example, can take up genes from other cells, or insert genes into them. This is one way that bacteria become resistant to antibiotics.

hypothesis a preliminary explanation for a phenomenon in nature that can be tested by observations or experiments

intelligent design the current name for a religious philosophy called "natural philosophy" in the early 19th century. The movement promotes the study of nature with the sole purpose of demonstrating that a particular type of God exists. Intelligent design claims to be scientific because it incorporates facts and bits of science, but its methods are completely different than those of modern science, especially in that it disregards facts that do not support its premise.

intron a DNA sequence which is found within the borders of a gene but which is removed during the process of making a messenger RNA because it does not contain protein-encoding information

invertebrate an animal such as an insect that does not have a spinal column

last universal common ancestor (LUCA) the most recent organism from which all current life on Earth descends

lipids fat molecules which have several functions in the cell, including forming membranes, storing energy, and passing signals

macroevolution evolutionary changes that happen over very long periods of time. This usually refers to the development of large new branches of life, such as vertebrates or mammals.

materialism the philosophy that living processes can be explained in terms of their physical and chemical properties. In evolution this implies that life could emerge through the normal action of physical and chemical laws on inorganic substances, without the help of an additional external force.

meme elements of culture, such as ideas, songs, or slogans, passed from one person to another in a process analogous to the spread of genes through a population

Mendel's laws the principles developed by Gregor Mendel to explain how heredity works in species that breed through sex. The principles are based on the fact that males and females equally contribute units of hereditary material of their offspring. The units that are passed along may be dominant or recessive, which explains why some features of an organism resemble those of the mother and others the father.

microevolution evolutionary changes that happen on a small scale, often within a single species, such as a change in the frequency of a particular allele within just a few generations

microRNA a small RNA molecule which does not encode proteins but docks onto other RNA molecules to influence whether proteins are made from them

microsatellite a short DNA sequence containing many smaller repeated sequences, found throughout genomes and used in DNA fingerprinting because of its tendency to evolve very quickly

microtubule a tubelike structure that is part of the cell cytoskeleton. It is made of subunits of the protein tubulin and has important structural functions in shaping cells, transporting molecules within them, and splitting pairs of chromosomes during cell division.

missing link a hypothetical "intermediate form" between two known species. Usually the term is used as a misinterpretation of the evolutionary idea that two species share a common ancestor. Thus, scientists do not expect to find fossils of a creature that is "half man, half ape"; instead, they are trying to define the genes and characteristics found in ancestors that gave rise to two or more types of descendants.

monochromatism   a very rare genetic disease in which people lack a gene needed to build functional cone cells in their eyes. People with monochromatism cannot distinguish colors and can only see shades of grey.

most recent common ancestor   the most recent organism from which two or more organisms descend

mutation   a change that happens in a gene when it is not perfectly copied. This usually involves the swap of one base pair for another, but is often used to refer to the insertion of extra nucleotides, or larger rearrangments of material in chromosomes.

natural selection   the process by which the characteristics of some members of a species make them more successful than others at passing their DNA on to the next generation. If the effect is very strong (for example, if one type of organism can survive and reproduce and others cannot at all), or if it lasts for many generations, the result is likely to be the development of a new species.

natural theology   a religious philosophy from the early 19th century which promoted the study of nature only to gather evidence that a particular type of God exists. All observations are interpreted in such a way to support this assumption, and contradictions that arise are attributed to the Devil or the imperfection of human perception or the mind. Natural theology was revived in the 20th century as a theology called "intelligent design," a movement which claims to be scientific because it incorporates facts and bits of science that support its assumptions.

Neanderthal man   a hominid species that lived in Europe and Western Asia from about 150,000 to 30,000 years ago. During some of this time Neanderthals coexisted with modern humans, but were not their ancestors.

negative eugenics   deliberate attempts to change the human species by preventing certain groups from reproducing. One method frequently used in the United States in the early 20th century was to forcibly sterilize people considered "unfit." In Nazi Germany the idea that these practices could improve the species was used to justify the Holocaust.

nucleotide   the basic chemical unit that makes up DNA and RNA molecules

ontogeny   an individual's biological development from a fertilized egg to adulthood

pangenesis   Darwin's flawed hypothesis about heredity, in which cells from various parts of the body produce particles called gemmules that collect in the sex organs and mix to create a new plant or animal

parthenogenesis   a type of reproduction in which an egg develops into an animal without being fertilized by a sperm. Only the mother's genetic material is used to create offspring. Parthenogenesis is a natural form of cloning found in plants such as blackberries and a few types of lizards, insects, and other animals.

path analysis   a statistical method developed by Sewall Wright to reveal causal connections in very complex situations

phenotype   the physical characteristics or behavior of a cell or organism. A phenotype develops flexibly through interactions between genes and the environment. The same genome can produce radically different phenotypes; for example, the complete set of human genes can build neurons or red blood cells.

photopigment   a protein or part of a protein whose structure changes when it is exposed to light, triggering events in cells that lead to vision and photosynthesis.

photoreceptor cell   a type of cell which uses photopigments to translate light into electrical stimulation. In complex animals these cells form the basis of eyes, and their signals are passed via nerves to the brain.

phylogeny   the study of the relationships between ancestral species and their descendants, or a tree diagram that shows these relationships

pistil   the female sex organs of a plant

plastid   organelles in plants which are responsible for photo-synthesis and the storage and manufacture of substances need-ed by the cell

polymerase chain reaction (PCR)   a method developed in the 1980s to make an unlimited number of copies (clones) of DNA molecules. Most experiments in molecular biology require millions or billions of copies of a molecule, so the development of PCR was a major advance in biotechnology.

population genetics   the quantitative study of the frequen-cy of genes and alleles in a population, and how their frequency changes over time

positive eugenics   a deliberate attempt to change the human species by encouraging specific people or groups to mate with one another

protein   a molecule made up of subunits called amino acids. Proteins are synthesized by cells using information in genes. They are often called the "worker molecules" of the cell be-cause of the many different functions they perform.

pseudogene   a DNA sequence that contained a gene (a protein-encoding sequence) in the past but which has un-dergone mutations and is no longer used for the synthesis of proteins

psychological isolation   behavior such as sexual preferenc-es that prevent members of the same species from mating with each other. Over the long term this can lead to the development of new species.

punctuated equilibrium a hypothesis developed by El-dredge and Gould stating that evolution has not always continued at an even pace. From an examination of the fossil record they claimed that there may be long periods in which species remain relatively stable, interrupted by short periods in which species change dramatically and many new ones are created.

recapitulation a hypothesis which holds that the development of an individual organism follows or repeats the long evolutionary development of its species

recessive allele an allele that has to be inherited from both parents in order to have an influence on an organism's development. A recessive allele has no effect (it is "silent") if it is inherited along with another allele that is dominant.

ribonucleic acid (RNA) a molecule made of nucleotide subunits that is made through the transcription of information contained in DNA. There are different types, including messenger RNA molecules, which are transported out of the cell nucleus and are used as patterns to build proteins. Other types are involved in building proteins, or in controlling whether other RNAs are used to do so.

scientific method a set of techniques to explore and explain natural phenomena based on collecting empirical data, formulating hypotheses, and testing them in rigorous, repeatable experiments

screening the practice of examining large numbers of organisms in the laboratory in search of mutations or other features

selective fertilization the idea that not every sperm can fertilize every egg, which could lead to reproductive isolation and the development of new species

selfish gene a view of natural selection from the perspective of the gene. A selfish gene is an allele that affects an organism in a way that makes it more likely to survive than other forms of the same gene.

sequence (DNA sequence)   a sequential list of the nucleotide bases that make up a region of DNA

sickle-cell anemia   a disease which is caused by mutations in a gene called globin. The defect causes the body to build red blood cells that are poor at transporting oxygen because they do not have the proper shape. The disease is dangerous, but the defective form of the gene is inherited at a high rate in some parts of the world where there are regular epidemics of malaria. (The parasite that causes malaria infects people by growing inside of healthy red blood cells.)

spandrel   a concept introduced by Gould and Lewontin that states that some features of organisms may be preserved through evolution without being subject to natural selection. The word comes from a technical term in architecture.

species   a group of organisms that are capable of breeding with each other to produce fertile offspring. Since new species usually develop gradually from a common ancestor, in some cases it is hard to tell whether two organisms belong to different species or are varieties of the same one.

spontaneous generation   a belief that fully formed organisms could arise from nonliving substances. Before scientists of the 19th century developed the cell theory, it was commonly believed that maggots or flies could arise by themselves, without eggs.

stamen   the male reproductive organ of a plant. A stamen usually has a pollen-bearing anther and a stalk.

stigma   a structure in a plant which receives pollen and serves as a female reproductive organ

survival of the fittest   a phrase invented by Herbert Spencer and reluctantly adopted by Charles Darwin to describe the result of natural selection. Although the term has become popular in discussions of evolution, it is misleading. The effect of natural selection is to change the gene pool of a species because some organisms reproduce more and pass along more of their genes to the next generation, and this is the only criterion for fitness.

temporal isolation a process by which different groups within a species stop producing offspring with each other because they mate at different seasons or different times of the day. Over time the groups may become different species.

theistic evolutionism the belief that evolution was the process by which God created life. In 2006 Pope Benedict XVI stated that this was his personal philosophy.

theory in science a collection of related hypotheses that can be tested through experiments and observations. To gain acceptance, theories usually have to be based on a large collection of data. Theories guide scientific work and help scientists interpret results, but they do not determine what counts as facts or evidence, so they can be overturned.

thermophile an organism which lives at high temperatures, often in hot springs or undersea thermal vents; many archaea and some species of bacteria are thermophiles

trilobite an extinct marine animal that lived from about 570–245 M.Y.A. Trilobites were arthropods, like insects and crustaceans.

tubulin a family of proteins made by eukaryotic cells and used to make microtubules, an important structural component of the cell cytoskeleton

variation the differences between individual members of the same species. Variation arises because in most species, organisms develop from unique mixtures of the genes of their parents. Mutations are another source of variation.

vertebrate all animals with a vertebral column made of cartilage or bone

vitalism a philosophy that holds that life cannot arise or be explained without a special "vital force" that animates nonliving matter

Further Resources

Books

Behe, Michael. *Darwin's Black Box: The Biochemical Challenge to Evolution.* Tenth Anniversary Edition. New York: Free Press, 2006. This book is a good example of the way current proponents of "intelligent design" attempt to enlist recent discoveries from science in support of a nonscientific hypothesis.

Blunt, Wilford. *Linnaeus: The Compleat Naturalist.* London: Frances Lincoln, 2004. A thorough and easy-to-read biography of the eccentric Swedish naturalist Carolus Linnaeus.

Browne, Janet. *Charles Darwin: The Power of Place.* New York: Knopf, 2002. The second volume of the "definitive" biography of Charles Darwin.

———. *Charles Darwin: Voyaging.* Princeton, N.J.: Princeton University Press, 1995. The first volume of the "definitive" biography of Charles Darwin.

Cadbury, Deborah. *The Dinosaur Hunters: A True Story of Scientific Rivalry and the Discovery of the Prehistoric World.* London: Fourth Estate, 2000. A popular account of the discovery, interpretation, and science of fossils in the 19th century.

Cameron, David W., and Colin P. Groves. *Bones, Stones and Molecules: "Out of Africa" and Human Origins.* Burlington, Mass.: Elsevier Academic Press, 2004. A very recent analysis of the hypothesis that modern humans originated in Africa, using fossil and genetic data.

Caporale, Lynn Helena. *Darwin in the Genome: Molecular Strategies in Biological Evolution.* New York: McGraw-Hill, 2003. A new look at variation and natural selection based on discoveries from the genomes of humans and other species, written by a noted biochemist.

Carlson, Elof Axel. *Mendel's Legacy: The Origin of Classical Genetics.* Cold Spring Harbor, N.Y.: Cold Spring Harbor Laboratory Press, 2004. An excellent, easy-to-read history of genetics from Mendel's work to the 1950s. Carlson explains the relationship between cell biology and genetics especially well.

———. *The Unfit: A History of a Bad Idea.* Cold Spring Harbor, N.Y.: Cold Spring Harbor Laboratory Press, 2001. An in-depth account of eugenics movements across the world.

Cavalli-Sforza, L. Luca, Paolo Menozzi, and Alberto Piazza. *The History and Geography of Human Genes.* Princeton, N.J.: Princeton University Press, 1994. This huge book studies the distribution of particular forms of genes from humans around the globe and shows what they reveal about how modern Homo sapiens settled the planet.

Chambers, Donald A. *DNA: The Double Helix: Perspective and Prospective at Forty Years.* New York: New York Academy of Sciences, 1995. A collection of historical papers from major figures involved in the discovery of DNA, with reminiscences from some of the authors.

Darwin, Charles. *The Descent of Man.* Amherst, N.Y.: Prometheus, 1998. In this book, originally published 12 years after *On the Origin of Species,* Darwin outlines his ideas on the place of human beings in evolutionary theory. All of Darwin's works are accessible to high-school level readers.

———. *On the Origin of Species.* Edison, N.J.: Castle Books, 2004. Darwin's masterpiece, outlining the full theory of evolution for the first time, and gathering a massive number of facts in support of it.

———. The *Voyage of the Beagle.* London: Penguin Books, 1989. Darwin's account of his five-year voyage around the globe, during which he collected specimens and made observations that would lead him to the theory of evolution.

Darwin, Charles, and Desmond King-Hele, ed. *Charles Darwin's "The Life of Erasmus Darwin."* Cambridge: Cambridge University Press, 2002. Darwin's account of the life and ideas of his grandfather, an early proponent of the idea that species change over time.

Dawkins, Richard. *The Blind Watchmaker.* London: Penguin Books, 1988. Dawkins's second book on evolution, in which he forcefully confronts the mentality behind "intelligent design" and argues that evolution is "the only known theory that could, in principle, solve the mystery of our existence."

———. *The Selfish Gene.* New York: Oxford University Press, 1989. Dawkins's first book on evolution, presenting evolutionary theory from the point of view of alleles competing for a "ride" on the next generation's genome.

Desmond, Adrian, and James R. Moore. *Darwin.* London: Penguin Books, 1992. An excellent biography of Charles Darwin.

Diamond, Jared. *Guns, Germs, and Steel: The Fates of Human Societies.* New York: W.W. Norton & Co., 1998. This is an extremely interesting study of human societies in terms of the interplay of environment, evolution, and culture.

Fara, Patricia. *Sex, Botany and Empire.* Cambridge: Icon Books, 2004. An interesting and entertaining account of the life and times of Joseph Banks, a naturalist, explorer, and politician, who determined the course of British science over several decades.

Gilbert, Scott. *Developmental Biology.* Sunderland, Mass.: Sinauer Associates, 1997. An excellent college-level text on all aspects of developmental biology, including "Evo-Devo" issues.

Goldsmith, Timothy H., and William F. Zimmermann. *Biology, Evolution, and Human Nature.* New York: Wiley, 2001. Life from the level of genes to human biology and behavior, from the point of view of evolutionary theory.

Gould, Stephen Jay. *The Panda's Thumb: More Reflections in Natural History.* New York: W.W. Norton & Company, 1992. A collection of essays on evolution and other topics.

———. *The Structure of Evolutionary Theory.* Harvard: Belknap Press, 2002. Gould's last book is an extremely detailed summary of 25 years of research, tracing the history of evolutionary thinking from Darwin to modern times.

Gregory, T. Ryan, ed. *The Evolution of the Genome.* Boston: Elsevier Academic Press, 2005. A detailed technical look at evolution from the perspective of how genomes have changed, written by 16 experts in various specialized areas of genome research.

Gribbin, John, and Mary Gribbin. *FitzRoy.* New Haven, Conn.: Yale University Press, 2004. The troubled life and times of the captain of the *Beagle.*

Hall, Michael N., and Patrick Linder, eds. *The Early Days of Yeast Genetics.* Cold Spring Harbor N.Y.: Cold Spring Harbor Laboratory Press, 1993. A collection of important papers from pioneers in the field of molecular genetics.

Hazelwood, Nick. *Savage: The Life and Times of Jemmy Button.* New York: Thomas Dunne Books, 2001. The story of a native of Tierra del Fuego, captured by Robert FitzRoy to be "civilized" in Great Britain and returned home during the second voyage of the *Beagle.*

Hazen, Robert. *Genesis: The Scientific Quest for Life's Origins.* Washington, D.C.: Joseph Henry Press, 2005. A scientist's personal investigation into new hypotheses about the origins of life on Earth.

Henig, Robin Marantz. *A Monk and Two Peas.* London: Weidenfeld & Nicolson, 2000. A popular, easy-to-read account of Gregor Mendel's work and its impact on later science.

Hooper, Judith. *Of Moths and Men.* New York: Norton 2002. The story of modern research into the "peppered moth" and its importance in evolutionary theory.

Jones, Stephen, Robert D. Martin, and David R. Pilbeam, eds. *The Cambridge Encyclopedia of Human Evolution.* Cambridge: Cambridge University Press, 1994. A collection of fascinating articles on all aspects of evolutionary theory up to the genome era.

Keynes, Richard Darwin. *Fossils, Finches and Fuegians: Charles Darwin's Adventures and Discoveries on the "Beagle."* New York: Oxford University Press, 2003. A retelling of the *Beagle's* voyage by a great-grandson of Charles Darwin, integrating material from Darwin's personal journals and notebooks.

Kirschner, Marc W. and John C. Gerhart. *The Plausibility of Life: Resolving Darwin's Dilemma.* New Haven, Conn.: Yale University Press, 2005. A new perspective on how mechanisms in cells create diversity in organisms—the material on which natural selection operates.

Kohler, Robert E. *Lords of the Fly: Drosophila Genetics and the Experimental Life.* Chicago: University of Chicago Press, 1994. The story of Thomas Hunt Morgan and his "disciples," whose discoveries regarding fruit fly genes dominated genetics in the first half of the 20th century.

Kohn, Marek. *A Reason for Everything: Natural Selection and the English Imagination.* London: Faber and Faber, 2004. This book focuses on the lives and work of researchers such as Haldane, Fisher, and Wright, who proved that genetic science was compatible with evolutionary theory during the first half of the 20th century.

Lutz, Peter L. *The Rise of Experimental Biology: An Illustrated History.* Totowa, N.J.: Human Press, 2002. A history of the development of modern biology, putting evolution into the context of what was going on in other parts of the field.

McElheny, Victor K. *Watson and DNA: Making a Scientific Revolution.* Cambridge, Mass.: Perseus, 2003. A retrospective on the work and life of James Watson, an extraordinary scientific personality.

Morris, Richard. *The Evolutionists: The Struggle for Darwin's Soul.* New York: W. H. Freeman and Company, 2001. An examination of several trends within current evolutionary theory, with a particular focus on Stephen Jay Gould and Richard Dawkins.

Purves, William K., David Sadava, Gordon H. Orians, and Craig Heller. *Life: The Science of Biology.* Kenndallville, Ind.: Sinauer Associates and W. H. Freeman, 2003. A comprehensive overview of themes from the life sciences.

Repcheck, Jack. *The Man Who Found Time.* London: Pocket Books, 2004. An account of the life of James Hutton and his impact on geology.

Ridley, Mark. *Evolution.* Oxford: Oxford University Press, 2003. An in-depth look at the modern science of evolution, probably best suited for entry-level college students.

Ruse, Michael. *Evolution Wars: A Guide to the Debates.* New Brunswick, N.J.: Rutgers University Press, 2001. A history of controversies that have surrounded evolution over the past 150 years.

Sacks, Oliver. *The Island of the Colorblind.* New York: A.A. Knopf, 1997. Noted author and psychologist Oliver Sacks's account of his trip to the island of Pingelap, many of whose citizens are afflicted by a curious genetic condition.

Severin, Tim. *The Spice Islands Voyage: The Quest for Alfred Wallace, the Man Who Shared Darwin's Discovery of Evolution.* New York: Carroll & Graf, 1997. Tim Severin retraces Wallace's route through the Spice Islands in modern times and offers a fresh new perspective on the scientist and his work.

Slotten, Ross A. *The Heretic in Darwin's Court: The Life of Alfred Russel Wallace.* New York: Columbia University Press, 2006. An excellent biography providing a particularly interesting look at the unusual course that Wallace's life took after the co-discovery of evolution.

Stearns, Steven, and Rolf Hoekstra. *Evolution.* New York: Oxford University Press, 2005. A very good, detailed introduction to the major themes in modern evolutionary research at a good level for an undergraduate introductory course to the field.

Stott, Rebecca. *Darwin and the Barnacle.* London: Faber & Faber, 2004. The story of Darwin's life and work in the decade between his discovery of evolution and his publication of the theory.

Tanford, Charles, and Jacqueline Reynolds. *Nature's Robots: A History of Proteins.* New York: Oxford University Press Inc., 2001. The story of the birth of biochemistry, giving a look at the lives and work of the researchers who uncovered the functions of proteins and other biological molecules.

Thompson, Harry. *This Thing of Darkness.* London: Headline Book Publishing, 2005. A novelized account of the voyage of the *Beagle,* focusing on the relationship between Darwin and FitzRoy.

Thomson, Keith. *The Watch on the Heath: Science and Religion before Darwin.* London: HarperCollins, 2005. An excellent historical overview and critique of "natural theology," the religious movement known today as "intelligent design."

Trinkhaus, Erik, and Pat Shipman. *The Neanderthals: Of Skeletons, Scientists, and Scandal.* New York: Vintage Books, 1994. A history of research into Neanderthals and the controversies within the field.

Tudge, Colin. *In Mendel's Footnotes.* London: Vintage, 2002. An excellent review of ideas and discoveries in genetics from Mendel's day to the 21st century.

———. *The Variety of Life: A Survey and a Celebration of All the Creatures that Have Ever Lived.* New York: Oxford University Press, 2000. A beautifully illustrated "tree of life," classifying and describing the spectrum of life on Earth.

Wallace, Alfred Russel. *Natural Selection and Tropical Nature: Essays on Descriptive and Theoretical Biology.* Boston: Adamant Media Corporation, 2005. A reprint of a collection of Wallace's major papers and writings, originally published in 1891.

Watson, James D. *The Double Helix.* New York: Atheneum, 1968. Watson's personal account of the discovery of the structure of DNA.

Wolpoff, Milford H., John Hawks, Brigitte Senut, Martin Pickford, and James Ahern. "An Ape or *the* Ape: Is the Toumaï Cranium TM 266 a Hominid?" *PaleoAnthropology* 2006: 36–50. A fascinating look at the way paleoanthropologists analyze teeth and bone structures to draw conclusions about how hominid fossils should be placed in the evolutionary tree.

Web Sites

There are tens of thousands of Web sites devoted to the topics of evolution, Charles Darwin, intelligent design, and the other themes of this book. The small selection below provides original articles, teaching materials, multimedia resources, and links to hundreds of other excellent sites. All sites listed were accessed June 1, 2008.

The American Society of Naturalists. "Evolution, Science, and Society: Evolutionary Biology and the National Research Agenda." Available online. URL: http://www.rci.rutgers.edu/~ecolevol/fulldoc.pdf. A document from the American Society of Naturalists and several other organizations, summarizing evolutionary theory and showing how it has contributed to other fields including health, agriculture, and the environmental sciences. Accessed on August 7, 2008.

Cold Spring Harbor Laboratory. "Image Archive on the American Eugenics Movement." Available online. URL: http://www.eugenicsarchive.org. An archive of images and material concerning eugenics from the Dolan DNA Learning Center of Cold Spring Harbor Laboratory, New York. The DNALC home page (www.dnalc.org) has many other excellent resources concerning biology, genomics, and health. Both accessed on August 7, 2008.

Cosmides, Leda, and John Tooby. "Evolutionary Psychology: A Primer." Available online. URL: http://www.psych.ucsb.edu/research/cep/primer.html. An excellent introduction to the field of evolutionary psychology by two of the foremost experts in the field. Accessed on August 7, 2008.

Estabrook, Arthur. "The Jukes in 1915." Available online. URL: http://www.disabilitymuseum.org/lib/docs/759.htm.   In 1875 Richard Dugdale carried out a study of 709 members of a family given the pseudonym "Jukes," trying to discover whether the abnormal number of criminals in the family had a genetic cause. His results were misinterpreted and used as propaganda for eugenics programs. In 1915, Arthur Estabrook, of the Carnegie Institution of Washington, did a follow-up study whose complete text can be found at the address above. Accessed on August 7, 2008.

Heslip, Steven. "Time-Space Chart of Hominid Fossils." Available online. URL: http://www.msu.edu/~heslipst/contents/ANP440/. A "comprehensive collection dealing with the history of hominid fossil finds throughout the world," listing the discoveries and the locations where they were found. Accessed on August 7, 2008.

The Institute of Human Origins. "Becoming Human." Available online. URL: http://www.becominghuman.org. A beautiful site with multimedia resources exploring human evolution. The Institute of Human Origins is a research organization affiliated with Arizona State University. Accessed on August 7, 2008.

Maddison, David R., and Katja Schulz, eds. "The Tree of Life Web Project." Available online. URL: http://tolweb.org. A site which has collected a huge number of articles and links from noted biologists on the question of assembling a "family tree" of life on Earth. Accessed on August 7, 2008.

Museum of Science, Boston. "Exploring Life's Origins." Available online. URL: http://exploringorigins.org. A beautiful multimedia site focusing on how life might have begun on Earth. Accessed on August 7, 2008.

National Academy of Sciences. "Evolution Resources from the National Academies." Available online. URL: http://www. nationalacademies.org/evolution. Evolution resources from the National Academy of Sciences, the National Academy of Engineering, the Institute of Medicine, and the National Research Council. There are several interesting books and articles for teachers that can be downloaded for free. Accessed on August 7, 2008.

The National Center for Biotechnology Information. "Bookshelf." Available online. URL: http://www.ncbi.nlm.nih.gov/ sites/entrz?db=books. A collection of excellent online books ranging from biochemistry and molecular biology to health topics. Most of the works are quite technical, but many include very accessible introductions to the topics. Some highlights are: *Molecular Biology of the Cell, Molecular Cell Biology,* and the *Wormbook.* There are also annual reports on health in the United States from the Centers for Disease Control and Prevention. Accessed on August 7, 2008.

National Center for Science Education. "Evolution/Creationism in the News." Available online. URL: http://www.natcenscied.org. The homepage of the National Center for Science Education, which collects stories in the news related to evolution and creationism. Accessed on August 7, 2008.

Social Science Research Council. "Race and Genomics." Available online. URL: http://raceandgenomics.ssrc.org. A Web forum in which experts discuss whether "human races" can be defined biologically in light of information from the human genome. Accessed on August 7, 2008.

Spencer, Stanley C. "Evolution Research News." Available online. URL: http://www.evolutionresearchnews.org. A collection of links to current articles and research publications on all themes related to evolution, assembled by biologist Stanley C. Spencer, intended as a research tool for biologists, students, and teachers. The site has an archive going back to 1997. Accessed on August 7, 2008.

TalkOrigins. "The Talk Origins Archive." Available online. URL: http://www.talkdesign.org. A Web site devoted to "assessing the claims of the Intelligent Design movement from the perspective of mainstream science; addressing the wider political, cultural, philosophical, moral, religious, and educational issues that have inspired the ID movement; and providing an archive of materials that critically examine the scientific claims of the ID movement." The site has several subsections, including the following: URL: http://www.talkdesign.org/cs/td_faq. An excellent summary of the main themes of proponents of Intelligent Design philosophies and their issues with evolutionary theory. URL: http://www.talkdesign.org/faqs/flagellum.html. A detailed discussion of author Michael Behe's claim that the bacterial flagellum is "irreducibly complex." Accessed on August 7, 2008.

University of California Museum of Paleontology. "Understanding Evolution." Available online. URL: http://evolution.berkeley.edu. An education Web site with excellent articles and materials for K-12 teachers, from the University of California Museum of Paleontology and the National Science Foundation. Accessed on August 8, 2008.

University of Cambridge. "The Complete Works of Charles Darwin Online." Available online. URL: http://darwin-online.org.uk. An online version of Darwin's complete publications, 20,000 private papers, and hundreds of supplementary works. Accessed on August 8, 2008.

The Vega Science Trust. "Scientists at Vega." Available online. URL: http://www.vega.org.uk/video/internal/15. Filmed interviews with some of the great figures in 20th-century and current science, including leading figures in evolutionary research. Accessed on August 7, 2008.

Index

Note: *Italic* page numbers indicate illustrations.